HumanBiology

breathing

P9-CLD-985

Text Authors

James V. Lawry
H. Craig Heller

Professional Library
Seattle Public Schools
Stanford Center 22-636

Activity Authors

James V. Lawry
Stan Ogren
Marjorie Gray
Geraldine Horsma

H. Craig Heller, **Principal Investigator**
Mary L. Kiely, **Project Director**

An Interdisciplinary Life Science Curriculum for the Middle Grades
Developed by the Program in Human Biology at Stanford University

EVERYDAY LEARNING®

Chicago, Illinois

Permissions

The Body Book by Sarah B. Stein, copyright © 1992 by Sara B. Stein. Used by permission of Workman Publishing Co., Inc., New York. All Rights Reserved.

Photo Credits

1 (top center), Tony Freeman/PhotoEdit; 8 (top center), Dennis O'Clair/Tony Stone Images; 19 (top center, Bill Aron/PhotoEdit; 28 (top center), National Aeronautics and Space Administration (NASA); 32 (bottom center), Paul Conklin/PhotoEdit; 33 (top center, AP/Wide World Photos; 34 (top center), Dave B. Fleetham/Visuals Unlimited; 38 (top center), John Livsey/Tony Stone Images; 39 (bottom center), © Biophoto Associates/Photo Researchers, Inc.; 43 (top center), CNRI/Science Photo Library (Photo Researchers); 44 (top center), © M. Abbey/Photo Researchers, Inc.; 45 (bottom left and right), UPI Photo.

Cover Image

SPL/Photo Researchers (x-ray of chest)

Everyday Learning Development Staff

Editorial

Steve Mico

Leslie Morrison

Susan Zeitner

Production/Design

Fran Brown

Annette Davis

Jess Schaal

Norma Underwood

Additional Credits

Project Editor: Dennis McKee

Shepherd, Inc.

ISBN 1-57039-679-5

Stanford University's Middle Grades Life Science Curriculum Project was supported by grants from the National Science Foundation, Carnegie Corporation of New York, and The David and Lucile Packard Foundation. The content of the Human Biology curriculum is the sole responsibility of Stanford University's Middle Grades Life Science Curriculum Project and does not necessarily reflect the views or opinions of the National Science Foundation, Carnegie Corporation of New York, or The David and Lucile Packard Foundation

Copyright © 1999 by Everyday Learning Corporation and The Board of Trustees of the Leland Stanford Junior University. All rights reserved. Printed in the United States of America. Individual purchasers of this book are hereby granted the right to make sufficient copies only of the reproducible masters in this book for use by all students in a single classroom. This permission is limited to a single classroom and does not apply to entire schools or school systems. Institutions purchasing this book should pass the permission on to a single classroom. Copying of this book or any of its parts for resale is prohibited.

Any questions regarding this policy should be addressed to:

Everyday Learning Corporation

P.O. Box 812960

Chicago, IL 60681

1 2 3 4 5 6 7 8 9 VL 03 02 01 00 99 98

Contents

Professional Library
Seattle Public Schools
Stanford Center 22-636

1

Breathing: Why and How?

What happens when I breathe?

Breathing moves air in and out of your **lungs**. Lungs are spongy tissue made up of many tiny air-filled sacs that are surrounded by blood vessels. Breathing moves air in and out of your lungs all your life. As you breathe in, your chest and lungs expand. As you breathe out, your chest and lungs contract (squeeze in). Look at the people around you. Can you see them breathe just by watching their chests move? Now observe your own breathing. Breathe in. How does your chest feel when you breathe in? How does your chest feel when you breathe out? This unit will help you learn about the breathing process—How air is moved in and out of the lungs—Why breathing is necessary—What breathing does for the body—And how the body knows how much and how fast to breathe. By the end of this unit it will be pretty obvious how important and how amazing your breathing system is. When you know how the breathing system works, you can explore how to keep your breathing system healthy.

Why Do You Breathe?

You breathe in to get the important gas called **oxygen** (O_2) from the air. All the cells in your body need oxygen. Without oxygen your cells, especially your brain cells, start to die in just a few minutes. You breathe out to get rid of a waste gas your cells make called **carbon dioxide** (CO_2).

apply your KNOWLEDGE A person who can't breathe will faint or "black out." What is fainting and why is it one of the first things to happen when you can't breathe?

How Do You Breathe?

Your lungs are enclosed in your chest cavity. Over the lungs are your shoulder bones. Around your lungs are your ribs. These bones together

Journal Writing

I Think Breathing . . .
Write a paragraph or a poem about breathing. Try to answer these questions in your writing. What is breathing? How can you tell if you are breathing or not? How does breathing look when you observe someone? How does it feel? How does it sound? What happens when you breathe more quickly? What happens when you breathe more slowly?

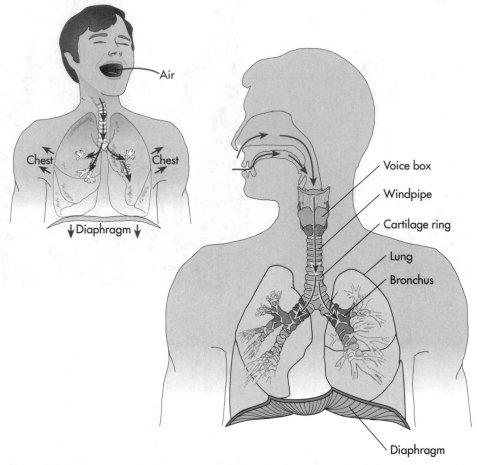

Figure 1.1 Air flows through the nose or mouth into the air passages. From there it flows to the air sacs in the lungs where oxygen and carbon dioxide are exchanged.

make the chest cavity stiff and strong to protect the lungs. The diaphragm is located beneath the lungs. The **diaphragm** is a large dome-shaped sheet of muscle. When the diaphragm muscle contracts, its dome shape is flattened and the chest cavity gets bigger. This change in the volume of the chest cavity pulls air into the lungs, and they expand. When the diaphragm relaxes, it moves back into its dome shape. When this happens, the chest cavity gets smaller and you breathe out.

As you breathe in, the air flows through your nose and mouth, through your windpipe, and through many smaller airways to all parts of your lungs. When you breathe out, the direction of the airflow reverses. So when you breathe out, the airflow goes from your lungs into the small airways, back up the windpipe, and out your nose or mouth.

apply your KNOWLEDGE

When you exercise, you breathe deeply. During exercise you take in more air per breath than when you are resting. Do you think this is because the diaphragm contracts more strongly when you exercise? What other muscles can help your diaphragm when you have to breathe deeply? Put your hands flat against the sides of your chest and breathe deeply. Can you feel other muscles helping you to breathe?

Let's make a model of your chest and lungs to see how your lungs inflate.

Activity 1-1
How Do You Breathe?

Introduction

How do you breathe? How does air move in and out of your lungs? In this activity you build a model of a lung to help you find out how the air moves into and out of your lungs. In the activity a clear, plastic soda bottle represents the chest. A balloon represents the lung. Another balloon represents the diaphragm or breathing muscle. The model you build will show how your lungs expand inside your chest each time you breathe. Your model also will show what happens if the breathing system is damaged by a wound that makes a hole in the chest cavity.

Materials

Clear, plastic bottle (about 1 liter)
Scissors
Tape
2 balloons, or one balloon and a rubber glove
Activity Report

Procedure

Step 1   Cut the bottom third off of the bottle. Be careful to leave a smooth edge so that the balloon won't tear.

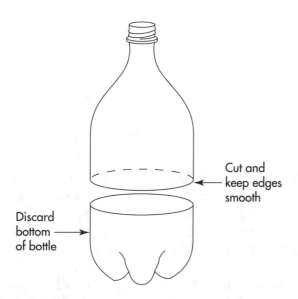

Figure 1.2 Cut and remove the bottom third of the bottle.

Step 2   Place the balloon inside the bottle and roll the open end of the balloon over the lip of the bottle opening.

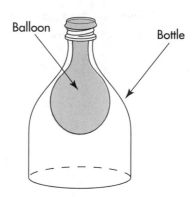

Figure 1.3 The balloon in the upper half of bottle should look like this.

Step 3   Tie off the neck of the second balloon. Cut the balloon so you can cover the opening at the bottom of the bottle with the balloon. Make sure the tied-off neck is facing down or outside the bottle after the balloon is attached.

Step 4   Tape the balloon to the bottle and make sure there are no air leaks.

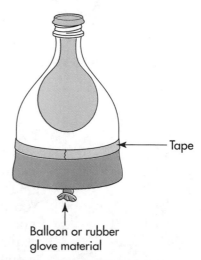

Figure 1.4 Cover the opening of the bottle bottom with another piece of balloon. Make sure the tied neck of the balloon faces out to use as a handle.

Activity 1-1 *(continued)*
How Do You Breathe?

Step 5   Use the tied off neck of the balloon as a handle and pull down. Hold for a second. Let go.

Step 6   Repeat. Observe and record what happens to the balloon inside the bottle.

Step 7   Answer questions 1–3 on the Activity Report before continuing with Step 8.

Step 8   Make a hole in the plastic chest. Pull down on the diaphragm. Hold for a second, and then release the diaphragm.

Step 9   Repeat. Observe and record what happens to the balloon inside the bottle.

Step 10   Answer questions 4–5 on the Activity Report. Make sure you save the model of the lung you built. You will be using the same model in Activity 5-2.

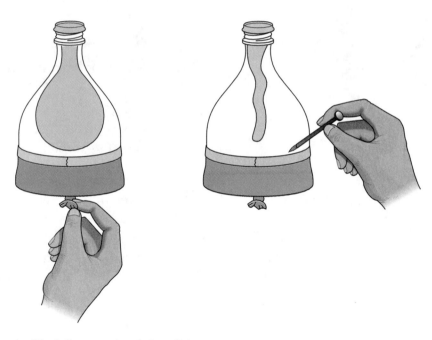

Figure 1.5 Use the tied off neck of the balloon as a handle to pull down.

Your lungs are similar in some ways to balloons. Like balloons, they fill with air. Your diaphragm is somewhat like a suction device. Air rushes into the lungs when the diaphragm contracts. Air is forced out of your lungs when the diaphragm relaxes. What makes your diaphragm work? You can make yourself breathe. But most of the time your diaphragm works, and you don't even know it. Your brain sends a message to the diaphragm muscle. The message tells the diaphragm to contract, and air is pulled into your lungs. The diaphragm moves automatically as part of your body's internal maintenance program.

Did You Know?

Lizards don't have diaphragms. All of their breathing depends on muscles between their ribs. These muscles are also needed for locomotion. So the lizards cannot breathe while they're running. That is one reason that lizards can be great sprinters, but not good long distance runners. Maybe the same was true of dinosaurs.

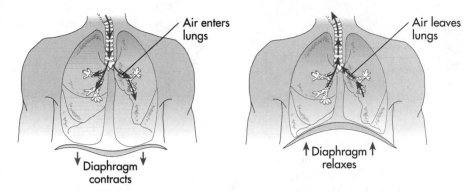

Figure 1.6 When the diaphragm muscle contracts and pulls down, air rushes into the lungs. When the diaphragm relaxes and moves up, air moves out of the lungs.

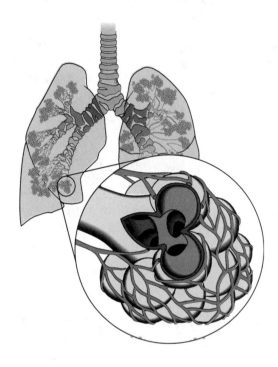

Figure 1.7 Your lungs aren't just two empty sacs filled with air like balloons. They're actually millions of tiny sacs bunched together.

apply your KNOWLEDGE

What are hiccups? What happens to your diaphragm when you get the hiccups? What causes hiccups?

What Do My Lungs Look Like?

Now try some creative thinking to imagine what it's like inside your lungs.

The inside of your lungs is pretty fantastic. Suppose you could shrink yourself down to a person smaller than the period at the end of this sentence. Now, suppose you enter someone's lungs in a breath of air. First, you bounce around and rattle down some tubes. As you get swept along, you pass many branch points like forks in a road. As you move through the tubes, they get smaller and smaller. Finally you end up

MINI ACTIVITY

How Many Breaths in a Lifetime?
Determine your resting breathing rate by counting how many breaths you take per minute while sitting at your desk. Then calculate how many breaths you will take during your lifetime if you live to be 85 years old.

How might exercise affect this number? Ask a partner to count the number of breaths you take after running or jumping in place for one minute. Then decide how many minutes you might exercise each week. Use what you just found out to recalculate your lifetime breaths.

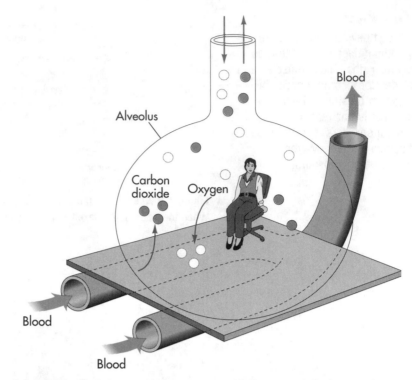

Alveolus

Blood

Carbon dioxide

Oxygen

Blood

Blood

Figure 1.8 A visit inside lungs would show you a world of nonstop action.

sitting at the bottom of a tiny, thin-walled sac. The only way to get out is the way you came in. This is true unless you can pass across the thin membranes of the tiny air sacs like molecules of oxygen can.

apply your KNOWLEDGE

Why would the Olympic cross-country, running team train at high altitudes?

Journal Writing

Write a poem or a song using the vocabulary in this section to describe what you think about breathing. You can relate your lungs to objects such as broccoli or relate your breathing to functions you commonly encounter such as wind outside your window.

Remember that you are still a tiny particle. Imagine that the floor of the air-sac you're in is bigger than the last tiny airway you came through. But the lining of the air sac is very thin—almost as thin as a soap bubble. You can see through it. You can see a lot of pipes under the thin, air sac floor. These pipes are the veins and arteries that carry blood to and from the heart. You hear blood rushing under you. You feel air whiz in and out with each breath. Oxygen gas pushes from the air around you, through the floor, and into the blood. Carbon dioxide gas bubbles up through the floor and out over your head. The walls of the air sac bounce in and out with every breath. It's very windy and noisy all around you.

Did You Know?
It wasn't until the 16th century that scientists actually began to test their theories about breathing in a systematic way. The theories of ancient thinkers such as Aristotle became the basis for practical experiments. In the mid-1600s, several scientists proposed that breathing actually supplied the body with a needed substance called air.

That imaginary trip helped show how you breathe. But there are some other important questions to answer. For example, "Why do you breathe? How do your lungs work? Where does the air go? What do you take out of the air you breathe in? What do you put into the air you breathe out? What does the air do in your chest? Why do you not pass out if you try to hold your breath?" These are some of the things you will learn in this unit. You'll also find out what exercising and climbing high mountains can do to your breathing. As you complete the unit, you'll learn about diseases that damage your ability to breathe. You will also find out how to keep your fantastic, breathing machine healthy.

Review Questions

1. Why do you need to breathe continuously?

2. Why do you exhale carbon dioxide (CO_2)?

3. Describe how your diaphragm works to make you breathe in and out.

2

My Breathing Machine

What do my lungs look like, and how do they work?

Remember that your lungs are not just balloon-like air sacs. The lungs are made up of millions of tiny air sacs. In fact, the lungs look like an upside down tree. The trachea is like the trunk that branches to your two lungs. In this section you will pretend to be a tiny explorer who explores the human breathing machine.

Now let's start to explore inside the breathing machine to see how air gets to the lungs. The airway to the lungs begins with the nose and mouth. Again, pretend that you are a tiny explorer only as big as the period at the end of this sentence. You begin your exploration at the nose. You crawl into a nostril and find yourself in a large chamber. It feels like the inside of a cave with wet, slippery walls.

The slimy walls are warm and moist. The slime is called **mucus.** The mucus is a secretion produced by the cells that line the airways. Mucus helps wet the air and traps dust and dirt to help keep the lungs clean. The curves in the walls of the nostril cave swish the air around to help moisten it. You don't always breathe through your nose. Sometimes you breathe through your mouth. When you do, the air goes straight into the pharynx. It doesn't swish around through curves. You may notice that breathing through your mouth, for even a short time, makes your mouth and throat feel dry and scratchy.

What Do You Think?

Cold medicines such as decongestants slow the production of mucus. As you learned, mucus is the body's way of catching and getting rid of germs and foreign materials. Do you think it is wise to take cold medications like decongestants? Why or why not?

Voice Box

Make a "guitar" from a cardboard box, a stick, and rubber bands. The stick keeps the bands away from the cardboard so they can vibrate. A thin, tight band makes a high sound. A thicker, looser band makes a low sound. Now blow across the bands. Describe the sound. What happens when you blow hard and when you blow lightly?

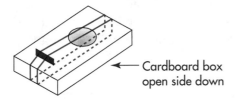

← Cardboard box open side down

Figure 2.2 Make a cardboard guitar like this one.

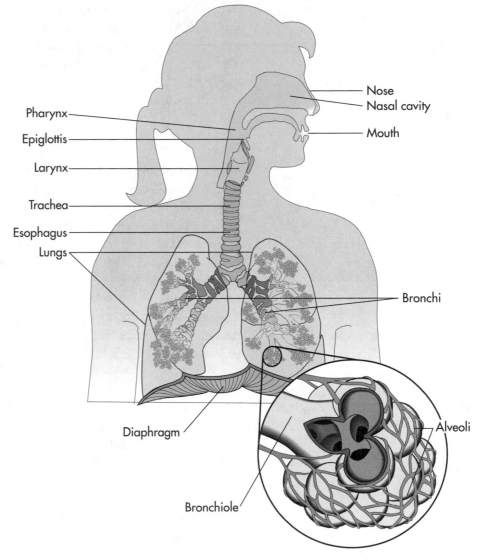

Figure 2.1 Your airways make it possible for your lungs to get lots of warm, moist, and clean air.

Did You Know?

Your head and chest act as the cardboard box does that you built in the Mini Activity. The chest and head like the cardboard box create a resonating chamber. The resonator moves with the vibrating string and makes a bigger sound. The shape of every person's skull and chest is different. So different people have different voices. No two human voices are exactly alike.

The nose passages and mouth cavity come together in the **pharynx.** The pharynx is the throat. The pharynx is the tube in back of the nose and mouth where the nose passages and mouth cavity meet. You crawl down the front side of the pharynx and come to a trapdoor that opens into a chamber strengthened with cartilage. You are now entering the voice box called the **larynx.** The larynx connects the pharynx with the windpipe that will take you to the lungs. You notice that there are bands of stiff tissue strung across the larynx. These bands of tissue are the vocal cords.

Let's find out how the vocal cords work. There is a slit-like opening called the **glottis** between the vocal cords. You draw your vocal cords apart when you breathe to let air into your windpipe. You draw the vocal cords together to speak or sing. You make sounds when you breathe out because the air flowing through the glottis causes the vocal cords to vibrate. The vibrating vocal cords make sounds like the vibrating strings on a guitar.

Now look at the side view of the head shown in Figure 2.1 on page 9. Find the tube just below the larynx or voice box. The tube that attaches to the larynx is the main breathing tube or windpipe called the **trachea**. The trachea is the main breathing tube to the lungs. Now look behind the trachea. There's another tube called the **esophagus**. The esophagus is the tube that carries what you eat and drink to your stomach. You can see that the trachea and esophagus are very close together. Also, they are both connected to the mouth and nasal passages.

So the esophagus and the trachea are both connected to the mouth cavity and nasal passages. But food doesn't go into the lungs. Let's find out why. To keep what you eat and drink from going into the trachea, a trap door sits over the larynx. The trap door closes when you eat or drink to keep food out of the airways. But it opens when you breathe in or cough out. This flap-like trap door is called the **epiglottis**. Notice the word *epiglottis* has the word *glottis* in it. The epiglottis is an extension of the glottis. Have you ever laughed so hard that milk you were drinking came out your nose? That happens when the epiglottis opened, letting air out of your lungs when you were drinking the milk and starting to laugh at the same time.

Did You Know?
When you hit a tennis ball really hard, it travels around 50 mph. When you hit a baseball really hard it travels about 85 mph. When you cough, particles and water bits can travel about 100 mph.

Sometimes food can get around the epiglottis and into your windpipe when you eat and talk or laugh at the same time. You cough when this happens. Food around your epiglottis triggers a cough. Then the blasts of air from the cough usually blow bits of food clear of your trachea. We will explore more about coughing and choking later in this unit. But the important concept to learn now is that the epiglottis covers your vocal cords and trachea when you swallow. That's why you can't swallow and talk at the same time. That's also why it's not a good idea to laugh when you have food or beverages in your mouth.

Did You Know?
When you get a really bad cold, your airways get infected. That infection is called *bronchitis* since it is the bronchi and bronchioles that are being attacked by the cold virus or bacteria.

Now let's get back to our exploration. By now you've gone past the epiglottis trap door and entered the trachea. You notice that the walls of the trachea contain rings of cartilage. Even from the outside you can feel the trachea in the front, low part of the neck. Below these rings of cartilage the trachea branches into two tubes—one tube for each lung. These tubes are called the **bronchi**. Like a branching tree, each of the bronchi branches again and again into smaller tubes called **bronchioles**. Each time the tube branches, it gets smaller. After about 16 forks in the path little clusters of air sacs called **alveoli** begin to appear.

apply your KNOWLEDGE   **What is the function of the rings of cartilage in your trachea?**

Activity 2-1
Building Clusters of Balloon Alveoli

Introduction
What do your lungs look like? In this activity each member of your class builds a balloon alveolus or air sac. Then you work with your team to build a balloon lung by clustering your individual balloon alveoli together. The airway into the alveolus is represented by yarn and the larger airway is represented by the rope.

Materials
Balloon
Yarn
String
Glue
Rope
Activity Report

Procedure
Step 1   Blow up a balloon. Each balloon represents an alveolus filled with air.

Step 2   Tie off the neck of your balloon with yarn so the air won't escape. Tie a piece of yarn to the neck of your balloon. The yarn represents the part of each airway that supplies the alveolus with air.

Step 3   Gather the balloons from all of the students in your group. Tie the free ends of the yarn together. Then tie the yarn ends to one end of the rope. (See Figure 2.3.) The rope represents a primary bronchus. The yarn represents the bronchioles. The balloons represent the alveolar sacs.

Step 4   Soak strings with glue. Then wrap the glue-soaked strings around the balloons. The strings represent small blood vessels around the lung called capillaries.

Step 5   Answer the questions on the Activity Report.

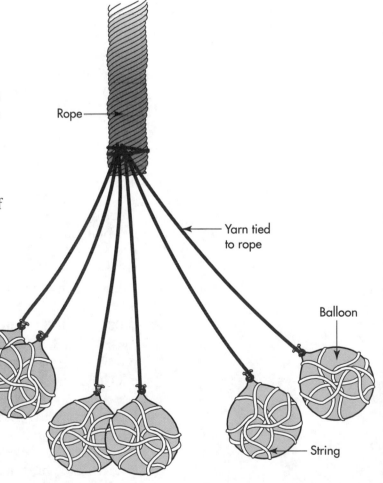

Figure 2.3 The balloons represent alveoli. The yarn represents small branches of the airways or bronchioles. The rope represents a bronchus. The strings represent the capillaries.

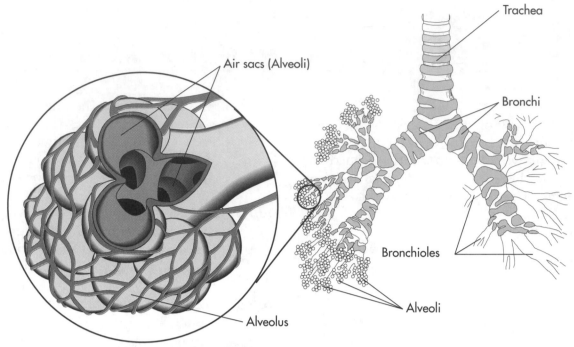

Figure 2.4 The airways called bronchioles branch to the air sacs called alveoli.

Alveoli

The alveoli are very important to breathing because that's where oxygen is taken in and carbon dioxide is released. To take in enough oxygen you need a lot of alveoli. Let's take a closer look at the alveoli to find out why we need so many.

How many air sacs or alveoli do you have? How large are the alveoli? First think about the tiny air tubes that open into the alveoli. Remember the airways branch many times, and each time they branch they get smaller. If your windpipe branches into two smaller tubes and each of them branches into two, how many branches would you have? Now if the branching happens twenty more times, how many branches would you have? It's a big number, isn't it? But the number of alveoli is even bigger because each final air tube opens into 60 or 70 alveoli. This is like a bunch of grapes on one stem. So how many alveoli do you have?

What do air sacs do?

Air sacs make it possible for your blood to get oxygen (O_2) from the air you breathe in. The alveoli also let your blood release carbon dioxide (CO_2) into the air you breathe out.

Blood that is low in oxygen and high in carbon dioxide comes from the heart to the lungs in big blood vessels called arteries. These **arteries** branch into smaller and smaller vessels called arterioles until they branch into the tiniest of all blood vessels called **capillaries**. These small capillaries surround each air sac or alveolus. Remember that the string you glued to the balloon alveoli in Activity 2-1 represented capillaries. The blood in the capillaries exchanges gasses through the thin walls of the air sacs. The gases pass through the thin walls of the

Did You Know?

Some animals such as certain kinds of frogs can exchange oxygen and carbon dioxide across their skin. They have thin, wet skin. Because they can obtain oxygen through their skin, they can stay under water for long periods of time.

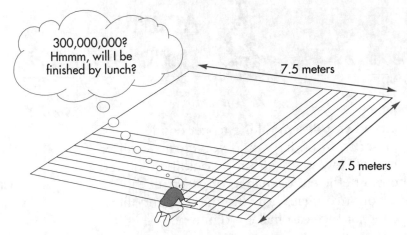

Figure 2.5 If you could count at the rate of 5 air sacs per second, it would take 694 days to count all the air sacs when they are squished-out flat— even if you counted 24 hours a day.

capillaries and the thin walls of the alveoli. This process is called **diffusion**. Diffusion is the natural movement of particles from an area of high concentration to an area of low concentration. For example, suppose there is a lot of carbon dioxide in the blood and only a little oxygen. But there is a lot of oxygen in the alveoli and only a little carbon dioxide. As a result of diffusion, the oxygen will move from the lungs where there is a lot into the blood where there is very little. The carbon dioxide will move from the blood where there is a lot into the alveoli where there is only a little. In this way, diffusion results in an even distribution of the particles.

apply your KNOWLEDGE

Different molecules diffuse at different rates. Give an example of different molecules that diffuse at different rates. Even one kind of molecule can diffuse at different rates depending on the conditions under which it is diffusing. Give an example of a molecule that can diffuse at a different rate when the temperature is cold than when the temperature is warm.

Did You Know?

Your body cells are never more than two cells away from a capillary. As the blood passes the cell, it brings food and oxygen and takes away carbon dioxide.

The small capillaries leave the alveoli and join into larger and larger blood vessels called **venules**. Eventually the venules join larger blood vessels called **veins**. The larger veins take blood that is high in oxygen (O_2) and low in carbon dioxide (CO_2) back to the heart. Then the heart pumps the oxygen-rich blood throughout the body to all the cells that need oxygen. You will learn more about why you need to exchange these gases in the next section.

Activity 2-2
The More the Airier

Introduction

How would you design a lung to get enough oxygen into your body? Would you create small or big air sacs? How many air sacs would be required for the most efficient exchange of gases? In this activity you investigate how the number and the size of the air sacs can make a difference.

Materials

1 volleyball
Ruler or tape measure (metric)
Calculator
4-liter freezer bag
16 Tennis balls
60 Golf balls
Activity Report

Procedure

Step 1 First, you're going to calculate the diameter squared (cm²) of different sized spheres. Begin by measuring the diameter of a volleyball. Next, measure the diameter of a tennis ball. Then measure the diameter of a golf ball. Record this information in Data Table 1 on your Activity Report. Then square the diameter number to give you the diameter squared. Record this information in Data Table 1 on your Activity Report. Note that the information for the alveoli is provided.

This is the equation you will use to calculate the diameter squared:

$$\text{diameter} \times \text{diameter} = \text{diameter}^2$$

We'll do one for you as an example. The diameter of a volleyball is 19 cm. To calculate its diameter squared, you use this equation:

$$19 \text{ cm} \times 19 \text{ cm} = 361 \text{ cm}^2$$

So the diameter squared of a volleyball is 361 cm².

Step 2 Now you are going to calculate the surface area for each type of ball. To calculate the surface area you will multiply the diameter squared of each ball by 3.1 (π). (π is the Greek letter pi. It represents the number you get if you divide the circumference of any circle by its diameter. Since this number is always the same, it is called a constant. The actual value of π is 3.1416. However for the purpose of the calculations in this activity we are using 3.1.) The equation would look like this.

$$D^2 \times \pi = SA$$

or

$$\text{diameter}^2 \times 3.1 = \text{surface area}$$

We calculated that the volleyball has a diameter squared of 361 cm². Now let's calculate the surface area of the same volleyball.

$$361 \text{ cm}^2 \times 3.1 = 1119.1 \text{ cm}^2 \text{ or } 1119 \text{ rounded out}$$

This is the surface area of the volleyball.

Now calculate the diameter squared, then the surface areas of the tennis ball and the golf ball. Record this information in Data Table 1 on your Activity Report. Then write the surface area of each golf ball in the appropriate column in Data Table 2 of your Activity Report.

Step 3 Your lungs can hold as much air as the 4-liter freezer bag can hold. How many volleyballs can be placed in the 4-liter bag? You do not need to squeeze the volleyball into a freezer bag to prove that only one volleyball will fit. Record your results in Data Table 2. Repeat this step for both tennis and golf balls.

Activity 2-2 *(continued)*

The More the Airier

Step 4 Calculate the total surface area of the lung volume by multiplying the surface area of one ball by the number of balls in the lung volume. You will use this equation to calculate the total surface area of the number of balls that fit in the 4-liter bag.

> surface area of ball × number of balls = total surface area

Now calculate the total surface area for one volleyball, since only one will fit into the bag.

> 1119 cm² × 1 Volleyball = 1119 cm²

Calculate the total surface area for the number of tennis balls that will actually fit

in the 4-liter bag. Then calculate the surface area for the number of golf balls that will fit into the 4-liter bag. Record this information in Data Table 2.

Step 5 Look at the total surface area of the alveoli in the lungs in Data Table 2. 60 m² is a little more than 7.5 m by 7.5 m. Use chalk or string to mark off an area 7.5 m by 7.5 m outside or on the classroom floor. This area represents the surface area of the lungs. Imagine all the alveoli are flattened out in this area.

Step 6 Answer the questions on the Activity Report.

MINI ACTIVITY

Words from the Latin Language
Research the word origins of the following words.
alveoli, glottis, and *trachea*

How Much Air Do You Breathe?

Activity 2-2 gave you an idea about how much air will fill your lungs. Let's see how much air your lungs really can hold. But before you measure the air that you move in and out of your lungs, consider that there is always some air in your lungs and airways. So if you breathe in a huge breath and then breathe out as much air as you could, would your lungs be completely empty? All of the airways starting with the ones with stiff rings of cartilage cannot be collapsed completely. Even the alveoli don't collapse completely when you breathe out. Now that you know that, here's the same question again. If you breathe in a huge breath and let it out, would your lungs be completely empty?

After you breathe out, there is always air left in your lungs and airways. There is a reason this last bit of air stays behind between breaths. Your breathing tubes, the trachea, bronchi, and bronchioles do not collapse when you breathe out. Your alveoli do not collapse entirely when you breathe out, either. Therefore, there is always some air in your lungs. The air left in your lungs between breaths does have some oxygen in it. However, that amount of oxygen wouldn't last long if you didn't breathe in fresh air again soon. Let's find out how much air you do breathe in and out.

Activity 2-3
Building and Using a Spirometer

Introduction

How much air do you breathe out each time you exhale? How much difference in volume is there between a normal breath and a deep breath? In this activity you build a spirometer to answer these questions. You use the spirometer to measure exhaled air after breathing normally and breathing deeply. Then you estimate how much gas you exhale in a normal breath and in a deep breath.

Materials

A large plastic container, basin, or washtub
   1/2 to 2/3 full of water
Two plastic gallon jugs
A funnel
Rubber tubing
100 ml (milliliter) beaker
Waterproof marker pens
Calculator
Plastic straw
Activity Report

Part A:
Procedure Build Your Spirometer

Step 1   Follow these steps to mark the volume on the jug.
- Fill the jug with water 500 ml (milliliters) at a time
- Mark the water level on the side of the jug for each 500-ml of water added.
- Number the 500-ml marks starting with the mark at the bottom of the jug.

Step 2   Place your straw into the end of the tubing. The straw will serve as a mouthpiece.

Step 3   Fill the basin or washtub half full of water. Fill the gallon jug so it is full of water. Invert the gallon jug in the basin while covering the opening so the water stays in the jug. If the jug doesn't stay up by itself, ask a partner to steady it for you. Then continue the procedure.

Step 4   Insert the end of the tubing that is opposite the straw into the opening of the gallon jug, and far enough in so the opening of the tube is close to the bottom of the jug, which is now on top.

Part B: Measuring Your Lung Volume
Measure your normal breaths.

Step 5   Take a normal breath. Breathe out through the straw into the hose as you would normally exhale. Don't force more air out than you inhaled.

Step 6   Record in Table 1 the volume of the air bubble at the top of the jug.

Step 7   Repeat and measure the volume 2 more times. Complete column 1 of the data Table 1. This is your **tidal volume.**

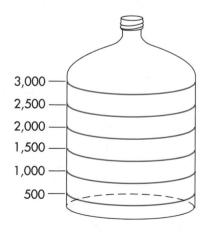

Figure 2.6 Mark each 500 ml on a jug.

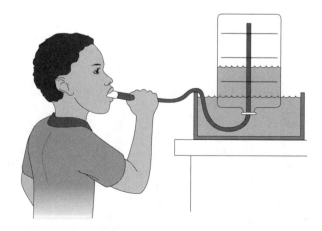

Figure 2.7 This drawing shows how a student uses a spirometer.

Step 8 **DISCARD YOUR STRAW** for sanitary reasons, so no one else uses it. Take a new straw for measuring deep breaths. Measure your rate of air exhaled for a normal breath.

Step 9 Calculate the average volume of air exhaled per minute. Record the rate in column 3 of Data Table 1. This calculation will give you an idea of how much air you normally breathe per minute.

Measure your deep breaths.

Step 10 Fill the jug full of water. Cover the mouth of the jug and invert in the basin. Take a deep breath. Force as much air out through the straw into the hose as possible. You may have to use two gallon jugs. When one is almost full of air, stop exhaling long enough to switch the breathing tube to a second inverted jug full of water.

Step 11 Record in Table 2 the volume of the air in the 1 or 2 jugs you used.

Step 12 Repeat and measure the volume two more times. This is your **vital capacity**. How does your vital capacity compare with the volume from a normal breath?

Step 13 **DISCARD YOUR STRAW** for sanitary reasons, so no one else uses it.

Measure your rate of breathing.

Step 14 With the help of your lab partner, determine how many times at rest you breathe in a minute. Record the rate in column 2 of Data Table 1. Repeat two more times. Record the rate in column 2 of Data Table 1. After averaging these results, you will have an idea of how much air you breathe in one minute.

Step 15 Measure your rate of air exhaled for a deep breath. Calculate the average volume of air exhaled per minute. Record the rate in column 3 of Data Table 2. This will give you an idea of how much air you breathe when taking a deep breath.

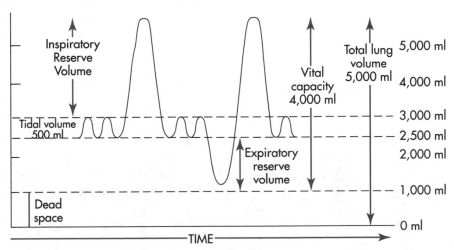

Figure 2.8 This graph shows lung volumes and capacities.

apply your KNOWLEDGE People who were born and live their whole lives in the high Andes Mountains of South America are described as being "barrel-chested." Why do you think this is so? What is the effect of altitude on the oxygen you breathe?

If you take a very deep breath and then exhale as much as you possibly can, the volume may be 6 to 8 times your tidal volume. This biggest possible breath is your **vital capacity.**

Use your spirometer to see how much of the difference between your vital capacity and your tidal volume is due to your ability to breathe more air in (your inspiratory reserve volume) or to breathe more air out (your expiratory reserve volume.)

Don't forget about the air in your lungs that you can never breathe out. That is your dead space. Your dead space may be as much as one liter. Therefore, your total lung volume is your vital capacity plus your dead space.

apply your
KNOWLEDGE

The air you breathe is 20% oxygen. Do you think the concentration of oxygen is the same in your alveoli when you breathe in? Explain.

Journal Writing

Pretend you are a small quantity of air. Describe your journey through your breathing system. Begin your journey entering the nose and mouth and stopping when you leave the breathing system to enter the blood capillaries. Your response can be in the form of a short story, cartoon, or poem.

Summary of the Parts of Your Breathing Machine

Your lungs are just one part of your breathing system. You have two lungs containing a total of 300 million alveoli (air sacs). If all 300 million air sacs in your lungs were squished out flat they would occupy a space about 7.5 m by 7.5 m. That's large enough to carpet half a tennis court. Calculate how many square meters there are in an area 7.5 m by 7.5 m.

The huge number of alveoli helps to increase the surface area that is available for gas exchange. But remember that there is more to your breathing system than only your lungs. In addition to your lungs, your breathing system includes your nose, larynx, trachea, and all the branches of your airways. And don't forget your diaphragm, and the muscles between your ribs that power breathing.

Review Questions

1. Explain how each major part of your respiratory system from your nose to your alveoli helps you to breathe.

2. How are your air sacs designed to maximize the amount of oxygen you can get from the air you breathe?

3. What happens in your lungs when you hold your breath?

3

Oxygen, Carbon Dioxide, and Energy

How are respiration and photosynthesis linked?

With each breath you exchange the gases oxygen (O_2) and carbon dioxide (CO_2). Why is this exchange important? Why does your body need oxygen? What happens to the CO_2? Where do we get O_2? These are some of the questions that this section will help you answer.

Every cell in your body needs oxygen to function. You get the oxygen your cells need from the air you breathe. The air you breathe is made up of 20 percent oxygen. The rest of the air is mostly nitrogen (79%). Your body cells use the oxygen you breathe to get energy from the food you eat. This process is called **cellular respiration**. During cellular respiration the cell uses oxygen to break down sugar. Breaking down sugar produces the energy your body needs. This is very similar to wood burning in a fire. As the wood burns, it combines with oxygen and releases heat energy and carbon dioxide. When the cell uses oxygen to break down sugar, oxygen is used, carbon dioxide is produced, and energy is released. But instead of heat energy, much of the energy produced in cellular respiration is stored chemically for the cell to use later. Carbon dioxide is the waste product of cellular respiration that you breathe out each time you breathe. Blood picks up oxygen and releases carbon dioxide in the lungs. The opposite takes place in the cells where the blood releases oxygen and picks up carbon dioxide.

The Breath of Life

Think about where the O_2 you inhale comes from and where the CO_2 you exhale goes. The exchange of gases doesn't only take place in the cells in your body. Actually, gas exchange is taking place all around you. In fact, oxygen and carbon dioxide are involved in the most important relationship that exists between plants and animals.

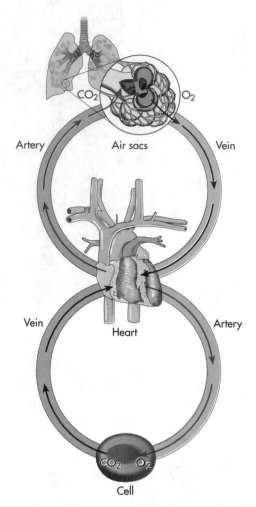

Figure 3.1 The blood in your lungs picks up oxygen and releases carbon dioxide. In your cells blood picks up carbon dioxide and releases oxygen.

Do you know where the oxygen your body needs comes from? Almost all the oxygen you breathe comes from green plants. They produce oxygen during a process called **photosynthesis**. During photosynthesis green plants manufacture the sugar molecules fructose and **glucose**. Green plants use energy from sunlight to build sugar molecules from carbon dioxide and water. Oxygen is produced when the plant combines the carbon dioxide and the water by using the Sun's energy. Plants use the sugar they produce to make plant structure and to provide the energy they need to live. Green plants use some of the oxygen they produce for their own life processes. But they release most of the oxygen produced during photosynthesis into the air as a waste product.

Unlike plants, which can get energy directly from sunlight, animals (including humans) must get energy from the food they eat. That food can be plants and/or other animals. Now think about the process of cellular respiration again. During cellular respiration animal cells combine oxygen with food molecules to release energy to live and function. Remember that cellular respiration produces carbon dioxide as a waste product. Animals use energy to grow, reproduce, and to function. They release the carbon dioxide into the air as a waste product. Plants help animals and animals help plants.

Did You Know?
It wasn't until the 1850s that scientists realized the Earth is a closed system—nothing comes in or goes out. Elements in our environment have cycles. They are used again and again. Recycling is a natural part of our existence. That's one reason we need to be careful and keep our environment clean and healthy.

Figure 3.2 Plants use carbon dioxide during photosynthesis to produce sugars and oxygen. Animals and plants use oxygen in respiration to produce carbon dioxide.

Respiration

All living organisms—plants and animals—carry out cellular respiration 24 hours a day. Cellular respiration takes place every minute of every day of every month of every year, and so on. The process of respiration can be shown with words or in a chemical equation. Both are written below.

$$\text{Oxygen} + \text{Glucose} \longrightarrow \text{Energy} + \text{Carbon dioxide} + \text{water}$$

$$6O_2 + C_6H_{12}O_6 \longrightarrow 36 \text{ ATP} + 6CO_2 + 6H_2O$$

Let's read the chemical equation for cellular respiration. The equation shows that cells use six molecules of oxygen ($6O_2$) to break down one molecule of glucose ($C_6H_{12}O_6$). The arrow shows that when the sugar is broken down something else is produced. Energy, water, and carbon dioxide are produced. When the chemical bonds of the glucose molecule are broken, energy is released. The cell is able to store that energy in the chemical bonds of a special molecule called adenosine triphosphate or ATP. The energy stored in ATP can be used by the cell to do various kinds of work. For each molecule of glucose broken down, 36 molecules of ATP can be produced. Finally when the glucose molecule is broken down, the waste products are six molecules of carbon dioxide ($6CO_2$) and six molecules of water ($6H_2O$).

Photosynthesis

The waste product of cellular respiration, carbon dioxide, is just what green plants need for photosynthesis. But green plants need more than just carbon dioxide to carry out the process of photosynthesis. They use a colored pigment called chlorophyll to collect energy from the Sun. During photosynthesis the plant uses the energy from the Sun to combine carbon dioxide and water to make the sugar called glucose. Like cellular respiration, the process of photosynthesis can be explained with words or in a chemical equation.

Did You Know?

One acre of trees produces enough oxygen to keep 18 people alive for one year. One tree absorbs about 13 pounds of carbon dioxide each year. One acre of trees absorbs 2.6 tons of CO_2. That's enough to offset the CO_2 produced by driving a car 26,000 miles.

MINI ACTIVITY

Miles = Trees

Estimate the number of miles that your family drives in one year. How many acres of trees would your family have to plant to offset the amount of carbon dioxide produced by the number of miles you estimated? What other ways, besides driving, cause CO_2 to be produced?

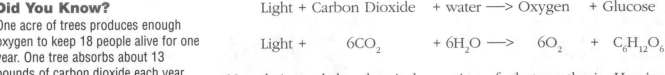

Light + Carbon Dioxide + water ---> Oxygen + Glucose

Light + $6CO_2$ + $6H_2O$ ---> $6O_2$ + $C_6H_{12}O_6$

Now let's read the chemical equation of photosynthesis. Here's what this chemical equation says. Green plant cells use the energy collected from sunlight to combine six carbon dioxide molecules ($6CO_2$) and 6 water molecules ($6H_2O$) to make one glucose molecule ($C_6H_{12}O_6$) and 6 molecules of oxygen ($6O_2$).

Now let's stop and think about the cycle we just investigated. Photosynthesis is the process by which plants use the energy of sunlight to make sugar (glucose) from carbon dioxide (CO_2) and water. Respiration is the process by which animals use O_2 to get energy from food. So plants take in CO_2 from the air and produce O_2. And animals take in O_2 from the air and produce CO_2.

Notice that these two chemical equations a related. Now we can do some activities based on the photosynthesis/respiration cycle. Photosynthesis and respiration form an extremely important cycle in nature.

apply your KNOWLEDGE Assume that all the oxygen you breathe in comes from plants and all the carbon dioxide plants use comes from animals. Why is the destruction of the tropical rain forest of concern to us in the United States?

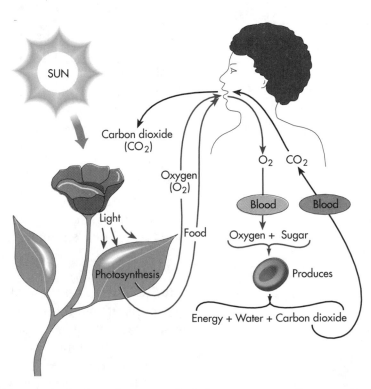

Figure 3.3 Plants and animals produce energy through reverse or opposite processes.

Activity 3-1
It's a Gas—Carbon Dioxide and Oxygen

Part A: The Gas—Carbon Dioxide
Introduction

When the cells of your body burn fuel (sugar), they produce a gas-carbon dioxide. When the cells of a green plant make sugar, they need a gas—carbon dioxide. In this activity you can see how carbon dioxide gets recycled. When you exhale, some of the carbon dioxide in your breath will dissolve in the water and turn the indicator bromthymol blue from blue to green to yellow. The indicator will change back to the original blue if carbon dioxide is not longer present.

Materials

Bromthymol blue indicator solution.
Plastic wrap or two test tube stoppers
Masking or labeling tape
Straws
2 Test tubes
Beaker or similar container
Elodea or some other water plant
Pond water
Activity Report

Procedure
Day 1:

Step 1 Write 1 on a small piece of masking tape. Write 2 on another small piece of masking tape. Place one of the pieces of labeled masking tape on one test tube and the other piece of labeled masking tape on a second test tube. Pour bromthymol blue solution into both of the test tubes labeled 1 and 2.

Step 2 Slowly bubble the air you exhale through a straw into each test tube.

Step 3 Observe and record your results on your Activity Report.

Step 4 Add sprigs of *Elodea* to test tube labeled 1 only.

Step 5 Cover the mouth of the test tubes with plastic wrap, wax paper, or a stopper.

Step 6 Place both test tubes in the light for 24 hours.

Day 2

Step 7 After 24 hours observe and record the results observed in the two test tubes.

Part B: The Gas-Oxygen

When the cells of a plant make sugar (fuel), they produce a gas—oxygen. In Part B of this activity you can see evidence of oxygen. By providing a strong light source for water plants, you can see the bubbles of gas the plant produces. This gas is oxygen—the gas you breathe and which our cells use to produce energy during cellular respiration.

Step 1 Place several healthy branches of *Elodea* in a large beaker of water.

Step 2 Place the beaker with the *Elodea* in a strong light source. Observe the *Elodea* and the water in the beaker over a 20-minute time period. Record your observations.

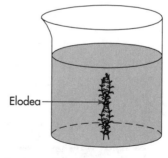

Figure 3.4 *Elodea* is shown in a beaker of water.

So you discovered that a green plant such as *Elodea* uses carbon dioxide and releases oxygen. That is opposite of what we and other animals do. Remember that we use oxygen and release carbon dioxide. As a matter of fact, your cells behave like candles in some ways. Both your cells and candles use oxygen to produce energy from fuel. Your cells use sugar as fuel. A candle burns the wax as fuel. Let's try another activity to see if candles use oxygen and release carbon dioxide like our cells do.

Activity 3-2
Cell Candles

Introduction

Do you know how candles behave like cells in your body? Both use oxygen to release energy from fuel. The candle burns wax. Your cells "burn" nutrients from food. The nutrient the cell uses for energy is glucose, a sugar, which is a carbohydrate. Both the cell and the candle produce carbon dioxide in the burning process.

$$fuel + 2\ O_2 \longrightarrow 2\ H_2O + CO_2 + energy$$

Let's read this equation. You know the word fuel. In your cells the fuel is glucose or sugar. In the candle the fuel is wax. So fuel combines with oxygen (O_2). The arrow means to become or produce. When fuel combines with oxygen, the combination produces water (H_2O), carbon dioxide (CO_2), and energy. Now try this activity to see if this equation really does happen when a candle burns.

Materials

1 Tall glass, glass jar, or beaker
1 Bowl larger than the glass, glass jar, or beaker
1 Candle
Matches
Water
Data Table
Activity Report

BE CAREFUL WITH MATCHES! Here's an important safety tip before you begin this activity. Make sure you are wearing safety goggles if you are using a match. Matches can be very dangerous, so be extremely careful. Make sure you douse the match and the candles in water before you discard them.

Procedure

Step 1 Carefully use a lit match to melt the bottom of a candle and stick the candle firmly in the center of the bowl.

Step 2 Fill the bowl with 3-cm of water.

Step 3 Carefully use a match to light the candle. Place a glass over the candle. Observe what happens to the candle, the water level, and the inside of the jar. Record your observations on your Activity Report.

Step 4 Repeat your experiment two more times. Record your data in your data table.

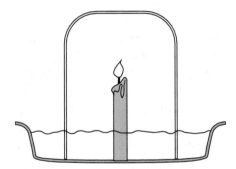

Figure 3.5 The bowl contains water with one candle covered.

 apply your KNOWLEDGE Explain how burning wood in a fireplace is like consuming glucose in your cells. How is it different? What is the chemical make up of wood?

Cellular Respiration and Burning Candles

Until the American Revolution people thought that air was just a single substance. A British scientist named Joseph Priestly showed that this couldn't be true. He showed that a candle in a sealed volume of air went out even though only a portion of the air was used up. He experimented with both a living animal, a mouse, and with a candle. He showed that introducing a growing, green plant could renew the oxygen in the air. His experiment showed that the mouse used up the oxygen in the air but not the nitrogen and other gases. And he showed that the green plant replaced the oxygen that was used up.

Let's look closer at the exchange of the two gases—oxygen and carbon dioxide. Remember that you can think of cellular respiration as being similar to a fire such as the burning candle. A fire uses oxygen to burn and gives off the waste product carbon dioxide (CO_2). Without oxygen a fire dies. When a fire doesn't have fuel it dies. When a fire burns fuel such as the wax in a candle, the molecules in the fuel combine with molecules of oxygen. When the molecules of oxygen and the molecules of fuel combine, the reaction releases energy as heat energy. Your body uses food as fuel to produce energy to keep you running all day. In a similar way, a candle burns wax, a car uses gas, or a train burns diesel fluid to keep going.

Did You Know?

Joseph Priestly's experiments were the first to demonstrate the cycle of respiration and photosynthesis. That's pretty interesting science information about Dr. Priestly. But here's a little history information about him. Joseph Priestly was an Englishman, but he supported America during the American Revolution. As a result, he became very unpopular in England and had to flee. He moved to Philadelphia where he became a chemistry professor at the University of Pennsylvania.

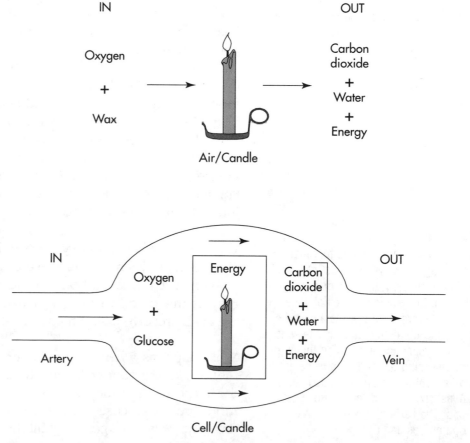

Figure 3.6 Cells are like candles. What did the candle under the glass do? It combined wax and oxygen, burned, and released carbon dioxide and water into the air. Wax is the fuel for the candle. Glucose (a sugar) is the fuel for your cells.

Just like a burning candle, your cells actually produce heat. How do you know your cells make heat? When you huddle under a blanket, you get warm. Why do you think you warm up under a blanket? Your body produces a lot of heat. Actually your body produces almost as much heat as a 50-watt light bulb does. But since the body is always making heat, you might ask this question. What does the blanket have to do with getting warmer? It's true that the body is constantly producing heat. However, a lot of the heat escapes through your skin. The more layers you cover yourself with, such as the blanket, the more heat is trapped. The blanket holds the heat close to you and keeps you warm. If you get too warm you remove the blanket to let the extra heat escape.

So animal and plant cells carry on respiration and green plant cells carry on photosynthesis. Now let's explore some other important topics about your breathing machine. For example, what happens to your breathing when you exercise?

Exercise and Breathing

Do your cells burn more fuel or glucose when you're resting or when you're playing soccer? Do you need more oxygen when you exercise? If you need more oxygen, then do you make more carbon dioxide? Let's find out.

What Do You Think?

Why do you think a person's breathing rate is one of the factors measured in a lie detector test?

Your heart pumps faster and your breathing rate increases when you exercise. Why do you think that statement is true? First imagine that cellular respiration is similar to a tiny candle burning in each cell. When you exercise, the tiny candles burn brighter. They burn their fuel faster and they use up the O_2 faster. As they burn the glucose fuel, they produce water and carbon dioxide. Your heart beats faster pumping more fuel and oxygen to the cells and taking carbon dioxide away from the cells. Your lungs inhale more and faster trying to take in enough O_2 and give off the excess carbon dioxide.

Figure 3.7 shows, in a simple way, how your heart and blood vessels connect to your cells and lungs. The heart pumps blood through the lungs. In the lungs the blood receives oxygen and releases carbon dioxide. Then the heart pumps this oxygen-rich blood from the lungs to your cells. When the oxygen-rich blood gets to the cells, the cells receive the oxygen and release the carbon dioxide. The blood with less oxygen and a lot of carbon dioxide returns to the heart. Then the heart returns this blood to the lungs where carbon dioxide is released and oxygen is received. And the Cycle starts all over again.

Did You Know?

Carbon dioxide is the gas that makes champagne fizz. When yeast cells use sugar for energy, they produce carbon dioxide. Sometimes bottles of wine are sealed before all of the yeast cells finish using up all the sugar in the wine. But the yeast continue using the sugar and releasing the carbon dioxide in the corked bottle. Carbon dioxide makes sodas fizz, too. But there's no yeast in soda. The carbon dioxide is forced into soda bottles and cans.

What happens when you exercise? The harder you exercise the more fuel or glucose your cells must burn. The more glucose your cells burn the more oxygen they use. So it's obvious that the cells need to get more fuel or glucose and oxygen to continue exercising. At the same time, the cells must get rid of all the carbon dioxide they are producing

Journal Writing

Write a letter to your town council in support of, or objecting to, a paper production company locating to your town. The company will bring lots of jobs to the area. But it will begin the cutting of the forest around your town, too. The people in your town need the jobs. But are the new jobs worth the cost of cutting down trees? Take into account the health effects of cutting the trees. Give the letter to your teacher to mail.

MINI ACTIVITY

Back to Normal

How long does it take for your heart and lungs to return to their normally functioning rate after you exercise? Find out. Find your heart rate while sitting quietly. Now run in place for 3 minutes. Take your heart rate again. Wait 1 minute and take your heart rate a third time. Wait another minute and take your heart rate again the fourth time. Record your heart rate each time. Continue to find your heart rate until it is back to your resting rate. How long did it take? Do you think this time period would get longer or shorter if you were an Olympic athlete who's been training for years? Now do the same experiment for your breathing rate. Try not to intentionally alter your breathing rate. It is best to have a partner measure your breathing rate while you think of other things.

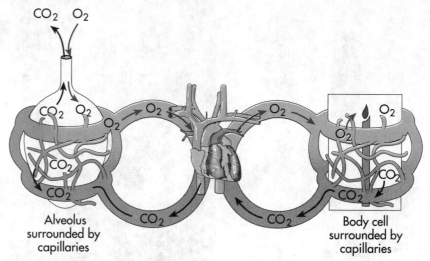

Figure 3.7 The heart is two pumps in one. One pump pushes oxygen-rich blood from the lungs to the cells of the body. The other pump pushes oxygen-poor blood back to the lungs.

as you continue to exercise. That's the reason your heart beats faster and your lungs move more air in and out when you exercise. Your heart and lungs are keeping up with the demands of your cells. Engineers admit that they couldn't design a more efficient system than the heart and lungs. Our hearts and lungs, like many other animals' hearts and lungs, work to match the flow of blood and gas to the needs of the cells. When the cells need more fuel and oxygen and need excess CO_2 removed, the heart and lungs work harder than when the cells need less.

Review Questions

1. Explain how photosynthesis and respiration are part of the same cycle. Draw a picture and describe the cycle.

2. Explain how your cells are like burning candles.

3. How do the cells in your body respond when you exercise?

4. How do your heart and lungs respond to the increased demand for oxygen when you exercise?

4

Breathing Mission Control

NASA Mission Control Center

How is breathing controlled to match the needs of the cells?

You breathe hard and fast when exercising. You breathe more quietly and slowly when you are sleeping. So how does your body know how much to breathe? Breathing is not usually a conscious decision. You can make yourself breathe faster or slower and sometimes hold your breath. But most of the time breathing happens automatically without having to think about it. The interesting question is, "How does your body know when to breathe and how hard to breathe?" How does it control oxygen intake and carbon dioxide release? In this section you will explore your body's **control systems**—the systems in your body that help regulate and maintain body functions at a normal level.

Controllers

Your body has to keep track of many things such as temperature, oxygen, energy, water, salt, nutrients, growth, balance, reproduction, and much more. Your body has control systems that maintain the functions in your body. Control systems detect changes from normal. When a change is detected, the controller sends messages to the organs of the body telling them to go into action to return conditions to normal. The controller for breathing is located in the parts of your brain just above the spinal cord. This area of the brain is called the brain stem. The breathing controller is within parts of the brain stem called the pons and the medulla. There are other controllers in the pons and the medulla, and there are many other controllers in the brain!

MINI ACTIVITY

Homeostasis

Look up the word *homeostasis*. What does it mean? Where does the word come from? How does it relate to feedback?

You will study a temperature controller in Activity 4-1 by controlling the temperature of a water bath. When you finish you will know more about how controllers use **feedback information** to keep something constant. Feedback systems can be based on positive or negative feedback. Positive feedback allows an action to continue and to get bigger and bigger. Negative feedback causes an action to stop or reverse. Most of the feedback systems in your body are negative feedback systems. For example, if you start to lose your balance and fall to the left, the balance detectors in your ears sense that you are falling to the left. They send this information to the balance controller in the brain. The balance controller sends instructions to your muscles to push you back to the right and restore your balance. Because the information from the sensors—falling to the left—caused an opposite response—move to the right—we say the sensors provided the controller with negative feedback information.

Another controller in the brain works like the thermostat in your home. A **thermostat** controls temperature. The thermostat in your home responds to changes in room temperature, and activates either the heater or air conditioner to keep the room temperature at a constant level. The thermostat senses the room temperature and compares the room temperature with a set point temperature. If the room is colder than the set point, the thermostat turns on the furnace. If the room is warmer than the set point, the thermostat turns off the furnace.

Your body has a controller that is similar to the thermostat in your home. Your body's thermostat is in a part of the brain called the hypothalamus. Your body temperature stays rather constant in all the seasons, at rest, and during exercise. Like a home thermostat controls the furnace, the body thermostat makes you shiver when you are too cold or sweat when you are too warm. The shivering heats you up and the sweating cools you down. A rise in body temperature causes the body's thermostat to activate responses that reverse the rise in temperature. Therefore, this controller is also using negative feedback information. Similarly, a fall in body temperature causes the controller to issue commands that reverse the fall in temperature.

apply your KNOWLEDGE Your breathing control systems use negative feedback. Can you think of any other negative feedback systems in your body?

Activity 4-1
How Does a Controller Work?

Introduction

There are many instruments that control variables and keep them constant. A good example is a thermostat that controls a water bath or furnace. In this activity you investigate how a controller works by becoming the thermostat. The purpose is to keep the temperature of the water at 37° Celsius (98.6° F). This is the normal temperature of your body.

Materials

Water bath; e.g. 1000-ml (milliliter) beaker
Thermometer
Ice cubes in container
Heater (hot plate)
Resources 1, 2, and 3
Activity Report

BE CAREFUL WITH THE HOT PLATE! Here's an important safety tip before you begin this activity. Make sure you are wearing safety goggles to protect yourself from hot water. Hot plates can be very dangerous. They get hot enough to burn. So be extremely careful.

Procedure

Step 1   Use Resource 1 as a model. There are four jobs for the members of your group to do. They are the following jobs.

Student 1 watches the thermometer and tells another student when to add ice or heat the water to keep the temperature at 37° Celsius.

Student 2 adds ice water to the bath if the thermometer reads greater than 37° Celsius.

Student 3 turns on the hot plate if the thermometer reads less than 37° C.

Student 4 records the temperatures and completes the temperature table.

Step 2   Design your own data table or use the Data Table on Resource 2.

Step 3   Heat the water in the beaker to a temperature of 37° Celsius. When the temperature is exactly 37° C, use this as time zero and start recording.

Step 4   Maintain this temperature at 37° C for 20 minutes. Adjust the temperature with ice or the heater to maintain at the temperature of 37° C. Take the temperature every 2 minutes and record it on the Data Sheet.

Step 5   Use Resource 3 to make a line graph showing the temperature changes during the twenty-minute period.

Controlling Your Breathing

Your respiratory system is responsible for the important exchange of the gases oxygen and carbon dioxide. Your breathing rate tells a lot about how much oxygen and carbon dioxide you have in your blood. The amount of carbon dioxide and oxygen in your blood is called your **blood gas levels**. The breathing controller in the pons and medulla of your brain receives information about how much oxygen (O_2) and carbon dioxide (CO_2) is in your blood. If the amount of CO_2 increases or the amount of O_2 decreases, your breathing controller makes you breathe faster and deeper. These same changes in blood gases cause your heart controller to speed up your heart rate.

The breathing controller works in a similar way as a thermostat works. The breathing controller compares existing levels of oxygen and carbon dioxide to set levels. Sensors pick up changes in levels of CO_2 and O_2 in the blood. The sensors send this information to the breathing controller in the pons and medulla. The breathing controller responds by sending instructions to the breathing muscles to breathe faster or slower and to the heart to beat faster or slower. For example, if there's too much carbon dioxide or too little oxygen, the lungs work harder and the heart pumps more.

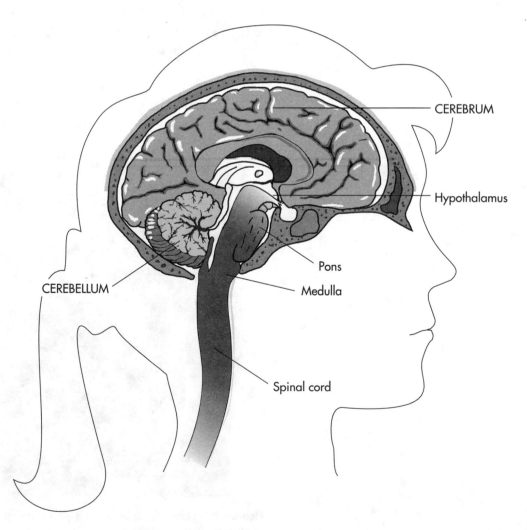

Figure 4.1 The medulla and pons control your breathing, and the hypothalamus controls your body temperature. Both systems use negative feedback.

Your Amazing, Adapting Body

Remember that when you exercise, your cells use more O_2 and produce more CO_2. That means the concentration of these gases in the blood changes. There is less O_2 and more CO_2. Your breathing controllers sense those changes in the amount or concentration of gases. When the breathing controllers sense the O_2 level in the blood is low and the CO_2 level is high, the controllers speed up your breathing and your heart rate. The increased breathing and blood pumping reverse the blood gas changes caused by exercise. Thus, the controller works to keep blood gases at the same level.

Sometimes the concentration of gases in your blood changes when the concentration of gases in the air you breathe changes. For example, have you ever climbed a high mountain? As you go higher and higher up a mountain there is less O_2 in the air. Since there is less O_2 in the air, the concentration or amount of O_2 in your blood is less, too. When this happens, you can feel "short of breath" and must breathe harder.

Mountain climbers that go to the tops of the highest mountains take oxygen with them in pressurized bottles. Gas is very compressible. This means gas can be forced into smaller containers than can the same amount of solids or liquids. So you can get lots of gas in a small bottle. What about you when you fly over those mountains in a jet plane? The airplane is pressurized to 5,000 feet. That means that being in an airplane is like being on top of a 5,000-foot mountain, no matter how high you fly.

Just as mountain climbers carry oxygen with them to the highest mountains, so do scuba divers carry their air supply in pressurized bottles or air tanks. When divers go very deep, their bodies are under high

Figure 4.2 At higher elevations you can quickly use up your blood's oxygen supply. That makes you breathe faster and harder.

Figure 4.3 Climbers on the world's highest peaks take oxygen with them.

pressure and they use the pressure in their air tanks to inflate their lungs. Breathing air at high pressure, however, causes gases to dissolve in their blood. Remember that as long as the bottle of champagne was corked the carbon dioxide stays dissolved in the liquid. But when the cork comes out, bubbles of gas come out of solution. If this happens in the blood of a diver who comes up too quickly, it is very serious. The gases dissolved in the diver's blood can come out of the blood as bubbles. This quick release of gas is called *the bends* and causes a lot of pain. The bends can be treated. A diver with the bends is put in a high-pressure tank to get the gases back into solution. Then the pressure is lowered slowly so the gas can diffuse out safely in the lungs.

Scuba divers carry their air in tanks that they wear on their backs. But some divers depend only on their lungs. Divers in the Far East who dive for pearls use only their lungs for air. They train themselves to be able to take very deep breaths of air that will last for several minutes. They decrease the O_2 they need for swimming by holding onto heavy weights to go down. Now suppose you are in the Far East diving for pearls. You take a big breath and dive. As you swim your body uses up oxygen and makes carbon dioxide all the time you are under water. The carbon dioxide is building up and the oxygen is being used up. Finally, the amounts of both of these gases tell your brain that you need to take a breath. That's when you are forced to come to the surface to breathe.

The breathing controller in your brain is more sensitive to increases in CO_2 in the blood than to decreases in O_2 in the blood. As you just read, sometimes people who want to swim long distances underwater spend a minute or so breathing fast and deep before beginning to hold their

Figure 4.4 Swimmers and divers often feel carbon dioxide build up and oxygen deplete if they stay under water too long.

breath. This is called *hyperventilation*. Hyperventilation does not add much O_2 to the blood, but it does eliminate a lot of CO_2. So when the swimmer starts swimming underwater, he or she doesn't feel the need to breathe and swims farther before surfacing. But this can be very dangerous because the swimmer can use so much blood O_2 that he or she can black out before feeling the urgent need to surface for air. Swimmers drown every year because they do not know how hyperventilation influences their breathing controllers.

As you can see, your breathing controller can adapt to different circumstances such as high pressure or low oxygen in the air for a short time. But your body also can adapt to changes around it for a longer period of time or even permanently. For example, people manage to live in high places on the top of high mountains. As a matter of fact, in spite of the low oxygen associated with living in high places nearly 15 million people in the world live over 10,000 feet above sea level. Some people live all their lives in the Andes Mountains in South America at 16,000 feet above sea level. How do people do this? Their bodies have adapted in several ways. One way the body adapts to less oxygen at higher altitudes is by increasing the capacity of the lungs. The lungs of people who live at high altitudes can often take in more air than the lungs of people at low altitudes. Another adaptation results in the production of more red blood cells. Red blood cells are the cells that carry O_2 to the cells. Another way the body adapts to altitude is by increasing the capacity of the blood to deliver O_2 to cells.

Did You Know?

One red blood cell contains 280 million molecules of hemoglobin. Each molecule of hemoglobin can carry 4 molecules of oxygen. How much oxygen can one red blood cell carry?

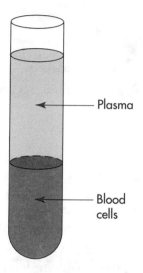

Figure 4.5 This test tube of whole blood demonstrates that your blood is made up of 40% blood cells and 60% plasma.

Plasma

Blood cells

Did You Know?

Your body makes 15 million new red blood cells every second to replace an equal number that die every second.

Your Hard-Working Red Blood Cells

Let's find out what's in your blood. Your blood contains water, salts, and blood cells. Most of the cells are red blood cells. Some of the blood cells are white blood cells. The white blood cells do several jobs such as fight infection. But the red blood cells are the cells that carry oxygen. Look at Figure 4.6 to see the shape of a red blood cell. It contains a protein called *hemoglobin*. Hemoglobin is able to pick up and combine with molecules of O_2 as the blood flows through the lungs. Then the hemoglobin releases that O_2 into the cells as the blood flows through the body and by the cells. You might say that the red blood cells are really containers filled with this wonderful protein. Red blood cells are red because the hemoglobin turns red when it combines with O_2.

Hemoglobin absorbs O_2 much like a sponge absorbs water. What happens when you whisk a damp sponge over a wet puddle on the floor? The sponge absorbs the water and the floor is dry. Now think of your lung model again. Imagine the blood filled capillaries coming close to the air filled air sac. Imagine that each red blood cell is like a sponge for O_2. When the red cell moves by the air sac, the O_2 leaves the air sac and goes into the red cell like water being absorbed by a sponge.

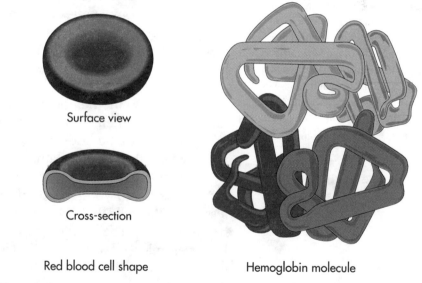

Surface view

Cross-section

Red blood cell shape

Hemoglobin molecule

Figure 4.6 O_2 binds to the protein in your red blood cells called hemoglobin.

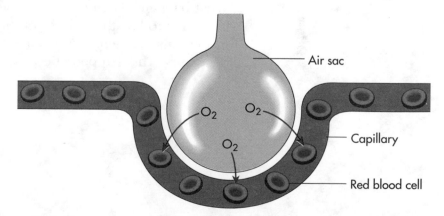

Air sac

O_2

O_2

O_2

Capillary

Red blood cell

Figure 4.7 Hemoglobin in red blood cells soaks up O_2 in your lungs and delivers it to your cells.

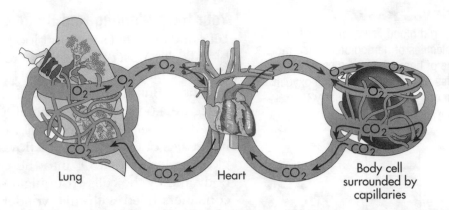

Lung   CO₂   Heart   CO₂   Body cell
                                                      surrounded by
                                                      capillaries

Figure 4.8 This drawing shows the path of oxygen and carbon dioxide. Red blood cells unload CO_2 and take up O_2 at the lung. They unload O_2 and take up CO_2 at the cell.

What Do You Think?

What is your opinion about the procedure that some athletes use called "blood doping?" Some athletes have some of their blood withdrawn and put in cold storage weeks before an event. Their bodies make new red blood cells to replace the ones that were withdrawn. The athletes have their stored blood transfused back into their bodies just before the event. This increases the number of red blood cells in their blood and the amount of oxygen they are able to take up from each breath. Why do you think an athlete might do this? Do you think blood doping should be an illegal procedure? Why or why not?

The red cell absorbs as much O_2 as it can the same way a damp sponge soaks up as much water as it can hold. When there is a lot of O_2 available in the air sac at sea level, each red cell takes as much as it can. But the O_2 in air sacs is lower at high levels such as on Mt. Everest. When the O_2 is low in the air sac, each blood cell picks up less than a full load. Therefore, these red cells have less O_2. So when they get to the body cells, there is less O_2 than is needed being delivered. To fix this problem, temporarily your body compensates by breathing faster and pumping more blood. Eventually your body will make more red blood cells if you stay at that elevation for a while. After a few weeks at high altitude, your body will have more red blood cells moving around in your blood than at sea level.

You know how the hemoglobin absorbs O_2 as the blood passes the air sac in the lung. Remember that the red blood cells loaded with O_2 are pumped back to the body and to the cells. Now let's find out what happens when the oxygen-saturated hemoglobin arrives at your cells. Remember that your cells are constantly using O_2 and making CO_2. The CO_2 diffuses into your blood and into your red cells. Because there is less O_2 in the cells than there was in the red blood cell, the hemoglobin releases its O_2. The O_2 diffuses into the cells that need it. When the O_2 has been released and the CO_2 is loaded, the blood takes the CO_2 back to the lungs. At the air sacs, the red blood cells release the CO_2 so you can breathe it out.

What happens when you hold your breath? Take a deep breath. Hold it. Try not to breathe. When you first breathe in, the amount of oxygen in the lungs is high. But as time passes, the O_2 content in your lungs decreases. This happens because the blood continues to take up O_2

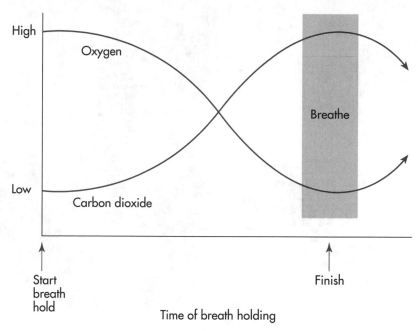

High

Oxygen

Breathe

Low

Carbon dioxide

Start
breath
hold

Finish

Time of breath holding

Figure 4.9 When you hold your breath, your body uses up the oxygen available in your lungs and begins to build up carbon dioxide levels.

from your lungs as it passes the air sac. But, since your mouth is shut and you are holding your breath, no new oxygen gets in to your lungs. As the O_2 is decreasing in your lungs the CO_2 is increasing. When you start holding your breath, the amount of carbon dioxide in your lungs is low. As more blood passes by the air sac, it takes away the O_2 but leaves the CO_2 it brought from the cells. So the amount of CO_2 increases in your lungs. Look at Figure 4.9. The graph shows the results of holding your breath.

Review Questions

1. How does the need for oxygen change as you exercise? How does the amount of oxygen in your blood change as you go to a higher altitude?

2. Describe how your body's controllers work like thermostats by using negative feedback.

3. How does your breathing controller work?

4. Describe how red blood cells and diffusion relate to breathing.

5

How to Keep My Breathing Machine Healthy

What are some common problems with breathing?

Your breathing system is amazing. Its shape and structure maximize oxygen (O_2) intake and carbon dioxide (CO_2) release. The breathing system also keeps out most elements that don't belong. How does your breathing machine help keep you healthy? How can you help keep your breathing machine healthy? This section will help you discover ways to keep your important breathing machine healthy and working efficiently for you.

Did You Know?

In ancient China, doctors were paid for keeping their patients healthy. They weren't paid if patients got sick. In fact, if a patient died, the doctor often had to pay the patient's family. In addition, for each patient that died, a special lantern was hung outside the doctor's office.

—David Louis, *2210 Fascinating Facts*.

Let's look at some of the main ways parts of your breathing machine keep you healthy. Two very important features that work nonstop are mucus and cilia. Mucus is a sticky secretion produced by cells in your breathing tubes. Little particles such as dust stick to the mucus instead of going into your lungs. **Cilia** are tiny hairs that line your airways. They collect tiny particles like the mucus does. But they do something else that helps get rid of the collected particles. They wave back and forth constantly to move things toward your throat. Their motion together with mucus creates a **mucus escalator.** Together the mucus and cilia create a moving sheet of mucus that brings particles up to your throat where you either swallow them or spit them out.

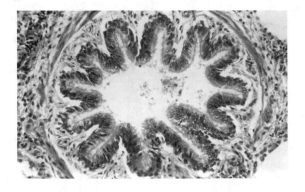

Figure 5.1 Once a particle sticks to mucus deep in your airway it is moved toward the throat. The particle rides along a moving sheet of mucus moved by the cilia, something like an escalator you might ride.

Illnesses

Your breathing system works remarkably well to keep foreign substances out to keep you healthy. But sometimes things do get into the breathing system and cause problems. Let's look at some illnesses that affect your breathing machine and some environmental factors that can harm your respiratory system.

Bacteria and viruses are so small that they can get by the mucus escalator. **Bacteria** are tiny organisms that can enter your body and interfere with the function of cells. Infections caused by bacteria are typically treated with antibiotics. **Viruses** are tiny living parasites. They are smaller organisms than bacteria are. Viruses also can cause illness but are not responsive to antibiotics. Even though antibiotics don't kill viruses, you can treat some of the symptoms they cause with medication. A major challenge to scientists and drug companies is to discover ways of fighting viral infections. What are some common illnesses caused by viruses and bacteria?

Flu and Colds

Flu, which is sometimes called influenza, and colds are caused by viruses. These viruses invade the cells of your airways. Then they begin reproducing and spreading like an invading army to other cells. Most cold viruses typically stay in your nose, throat, and chest areas. But a flu virus can spread throughout your body.

Colds and flu affect your airways greatly. They cause your cells to produce more mucus than you can effectively cough up or swallow. You feel congested as mucus fills your airways. The good news is you'll feel better in a few weeks unless bacteria begin to grow in the airways the viruses invaded. The bad news is you feel terrible for a week or more. The flu can affect your whole body with fever, aches, and pains. In severe cases, it can take a person as long as a month to recover from the flu.

Do you know what the word contagious means? Have you ever had a contagious illness? You probably answered yes since colds and the flu are contagious illnesses. Once they start to spread, many people in a community can become infected. Have you ever heard the phrase, "the common cold?" That means that colds are not specific. They are pretty much the same and have pretty much the same symptoms. For example, common symptoms include a stuffed up or runny nose, sneezing, watery eyes, coughing, and sometimes a sore throat. So cold symptoms are very common. But flu viruses are very distinct. They can have different types of symptoms. Some flu viruses cause muscle aches and pains. Some cause respiratory problems. Some cause digestive problems such as vomiting or diarrhea. Some can cause fevers and headaches. And some cause all of these symptoms. Flu can spread from region to region, which is called an epidemic. Or a particular strain of flu can spread worldwide, which is called a pandemic. You may have heard of the Asian Flu or the China Flu. Both of these types of flu were named because of where they began or were first discovered. Flu epidemics last four to six weeks and occur usually in winter.

Did You Know?

"There are all together billions of species of microbes that might, with the right combination of mutations, infect humans. But the chance of hitting on a workable combination is much less than a person's chance of choosing the right combination of numbers to win the California lottery. Of the billions, fewer than 500 germs have been lucky enough to have won the prize of infecting humans."
—excerpted from Sara Stein's *The Body Book*.

Journal Writing

What does having a cold feel like? How is your breathing affected? How do your lungs and your nose feel? What do you do to make yourself feel better? Do you think it is a good or bad idea to exercise when you have a flu or a cold? What can you do to prevent future colds?

Did You Know?

Influenza can be a very serious disease in older people and children. In 1957, the Asian flu killed 70,000 Americans. However, tens of millions caught the flu and recovered. More than half a million Americans died during the influenza epidemic of 1918.

Did You Know?

What are researchers doing to combat the flu? They are trying to find better ways to treat and prevent colds and flu.

- Biochemists are working on chemicals that cause a person's immune system to make a protein called interferon, which slows the growth and spread of viruses.
- Pharmaceutical chemists are making nose drop and spray vaccines that work faster than shots since they can be sprayed directly into your airways and nose.

Figure 5.2 You can get a virus from sharing towels, telephones, and other items used by a person with a viral infection. Many virus particles can live a short time outside of a living human.

 apply your KNOWLEDGE **How do you think flu epidemics can spread around the world in only a few weeks?**

You've probably been told that you should cover your mouth when you cough or sneeze. There's a good reason for that advice. When you cough or sneeze, you send millions of particles from your mouth and nose into the air. If those particles are viruses or bacteria that have infected you, then someone else may breathe what you just sneezed out. But there are other ways you can get flu and cold viruses, as well as bacteria. Viruses and bacteria can be spread on towels, telephones, toothpaste tubes, eating utensils, and dishes that were used by someone who is sick. Different viruses live for different amounts of time outside of a living human. Unfortunately you can't kill a virus with medicine. But you can treat cold and flu symptoms to make you feel more comfortable until the virus dies or becomes inactive.

You can lower your chances of catching cold or flu viruses in several ways. Here are a few of the ways you can help your body fight off the viruses and bacteria that are all around you.

1. Avoid physical contact with sick people or their belongings.
2. Wash your hands often.

It may seem like very simple advice to wash your hands. But the simple act of washing hands has decreased human mortality and increased human life span. At the time of the American Revolution, doctors and even surgeons did not wash their hands before or between the treatment of patients. The simple practice of washing hands with antiseptic solutions did more to reduce the death rate in hospitals than did the discovery of antibiotics.

3. Drink a lot of water and other healthy fluids, such as fruit juices. Drinking liquids helps because liquids are needed to keep your mucus wet and mobile. Mucus traps and removes viruses the same way it does particles in air pollution.
4. Make sure you get enough sleep to stay rested.
5. Avoid excessive stress and pressures.
6. Don't smoke. Smoking damages the airways. It makes the mucus lining thinner and it paralyzes the cilia that move the mucus toward the mouth. So smoking makes it easier for viruses and bacteria to get into the lungs.

Flu Vaccines

A flu vaccine is either a noninfectious form of the virus grown in eggs or a vaccine created in a test tube by genetic engineering. A flu shot may produce some mild symptoms that you would experience if you actually caught the flu. But flu shots don't cause flu. Nevertheless it makes the person's body produce antibodies that can attack the same virus if one comes along. Vaccines can usually protect against only two or three viruses. Unfortunately, viruses change rapidly as they spread through communities of people. Even so, doctors recommend that the elderly and individuals with chronic illness get flu shots. By following the origins and the spread of flu viruses, scientists try to produce vaccines each year that will be effective against the most common flu viruses. Scientists at the Center for Disease Control in Atlanta, Georgia, track flu outbreaks and predict which strains will be the biggest problem so the right vaccines can be produced. Unfortunately, so far there is no vaccine for the common cold.

Scientists have recently exhumed (dug up) bodies of flu victims that were buried in frozen soil in Alaska in 1918. From tissues of these bodies they have isolated the virus that caused the disease. They can now make and store a vaccine in case such a killer virus reappears.

Pneumonia and Bronchitis

Sometimes pneumonia or bronchitis can develop from a cold or flu virus. Colds and flu can challenge your breathing system so much that your cells can't effectively fight off infections.

Pneumonia is an infection in the lungs. Pneumonia occurs when the small air tubes and air sacs in your lungs get infected with bacteria. As the bacteria infect the cells, fluids are produced. The fluids from the infection begin to leak from the infected cells into the air sacs and airways. This fluid keeps enough oxygen from getting into the air sacs and into your blood to make you short of breath. As more fluid fills the air sacs and airways, it becomes even harder for you to breathe. The reaction of your body may be a high fever and a persistent cough. Your body causes you to cough to get rid of the fluids and the bacteria. Anybody who is short of breath resulting from the flu needs to see a doctor. Pneumonia usually requires medical treatment for the patient to get better. People have died from pneumonia because the fluids continue to fill the air sacs so the lungs eventually get no oxygen. This process is similar to drowning.

Bronchitis is an infection of the upper airways and smaller air tubes. A persistent cough and mild wheezing are two symptoms of bronchitis. The air sacs and airways are not infected. So gas exchange is fairly normal and patients with bronchitis are not usually short of breath. Bronchitis often needs medical treatment for the patient to get better. Some smokers have chronic bronchitis. The word chronic means a constant condition that never goes away. So many smokers have chronic bronchitis that never goes away. The smoke irritates the lining of small airways and stops the mucus escalator. Because the airways accumulate mucus and debris the smoker has a hacking cough. The more debris and mucus that collects the worse the smoker's cough gets. Another problem caused by smoking results from the destruction of the cilia and mucus cells. With fewer cilia and mucus cells it's much easier for smokers to contract pneumonia and bronchitis than it is for nonsmokers.

Wheezing and Asthma

Many people are asthmatic. Anyone who is asthmatic has some difficulty breathing. **Asthma** is a condition in which the small airways leading to the air sacs get narrow. Air moving through the smaller tubes can make a wheezy sound. Mucus glands in asthmatics may make more mucus. More mucus causes people to cough and increases their difficulty in breathing. About three-quarters of asthma victims are allergic to dust and/or bacteria. People with asthma need to drink lots of water. Drinking a lot of water helps to thin the mucus. Some asthmatics need some medication to open their airways.

Tuberculosis, an Old Threat That's Returned

Have you ever heard of **tuberculosis** or TB? You don't hear much about TB in the United States anymore. But TB is still one of the ten leading causes of death in the world. TB may still be a big problem in the United States for several reasons. People coming into the US from other countries can bring TB with them without showing any symptoms of the disease. On top of that, new TB germs have developed that do not respond to modern medicine. All these factors make TB a continuing threat to world health.

Did You Know?
About 3 million people die of TB each year. It is the major infectious disease in the world. Almost 2 billion people carry TB bacteria. But in most people, the bacteria are inactive.

TB is a bacterial disease. You can breathe the bacteria into your lungs if you are around someone with TB who sneezes, coughs, or speaks. The scary thing about TB is that it is a silent disease. In other words, you can be infected with TB without showing any symptoms of the disease for many years. You can get TB from people who don't know they have it, and you can have TB and not know it. Also, you can become infected with TB and immediately show symptoms. Those are the reasons people get tested for TB.

Did You Know?
The word *bacteria* is plural. The word *bacterium* is singular. So you can have millions of bacteria in your environment. But you might see only one bacterium under your microscope.

You cannot spread TB unless the bacteria are active in your body. When the bacteria are active, the tuberculosis bacteria destroy healthy lung cells and groups of cells called tissue. The body replaces the destroyed tissue with scar tissue. Scar tissue is thick and not very elastic. It can't expand or contract as air flows in and out. Since scar tissue is thick, it prevents gases from diffusing to and from the blood as easily as they can in a healthy air sac.

Did You Know?
Have you ever had a TB test done as part of your regular checkup? You've probably been tested in one of two ways. One test uses a four-pronged skin prick and is called the *TB tine test*. Another test involves a simple injection just under the skin. Both methods detect if you have ever been infected with TB bacteria.

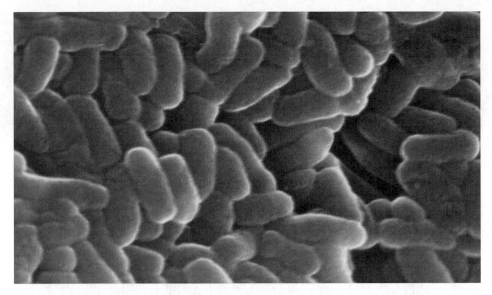

Figure 5.3 Active tuberculosis bacteria like those shown in this picture destroy the air sacs of a person's lungs.

apply your KNOWLEDGE

Okay, you learned that TB can't be spread unless the bacteria are active. Then what is the significance of TB being a silent disease? If you can carry the disease without becoming sick then the bacteria in you are not active.

Environmental Causes of Breathing Problems

Air pollution

Air pollution is a name for the substances put into the air that contaminate our atmosphere. Ash, smoke, and exhaust from automobiles are types of air pollution. These kinds of pollutants make your breathing system work harder than usual to keep you healthy.

Air pollutants can be divided into two forms. Particulates are small particles of solids such as particles of dirt, ash, and soot. These particulates float around in the air. The second form of pollution occurs when these particulates get trapped in the air close to the ground. Normally heat from the ground warms the lower layers of air near the ground. Then the warmer layers of air expand, become less dense, and begin to rise. The rising air can carry away pollution. But if the ground is cold or cool, the air stays dense and doesn't rise. This situation is called a temperature inversion. In bad cases, thick smog can result. Many chemicals remain suspended in the air like a mist or a fog. This mixture is called an **aerosol**.

When you inhale an aerosol, big particles such as dust and soot swirl around in the nasal passages. They land on the wet mucus in the nasal passages and are trapped. Smaller particles can get to the airways and land on the mucus there. And some very small particles can get into the air sacs. Some small particles may never land at all, but are exhaled with the next breath.

Air pollution, which includes cigarette, cigar, and pipe smoking, can cause lung cancer. Lung cancer causes cells in the lung tissue to behave

MINI ACTIVITY

Teens and adults smoke for many reasons. Often people make excuses for what they do. How many of these have you heard?

• Adults smoke, so why shouldn't I?
• One cigarette never killed anybody.
• Life's tough. I smoke to cope with the pressure.
• I just want to be part of the group.
• Smoking is cool.
• I smoke to rebel.
• I smoke so I won't overeat and get fat.

Write a good argument against each of these statements.

MINI ACTIVITY

Take Action! Write a letter and have your teacher mail it to an elected representative stating your position on some aspect of air pollution. Here are some examples to get you started thinking—smoking, factory emissions, or car emissions.

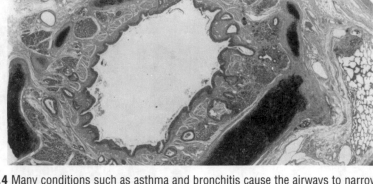

Figure 5.4 Many conditions such as asthma and bronchitis cause the airways to narrow. The narrower airways then make it more difficult to breathe. This cross-section of a bronchiole shows an airway (light area, center) surrounded by mucus.

abnormally to reproduce without limit. Growths of abnormal cells called tumors can clog airways. Cancerous cells may overproduce mucus or slow, even stop, its movement up the airways. When the mucus becomes trapped in the bronchial tubes, a cough develops.

Cancer is not the only possible harmful effect of air pollution, though. Even without causing cancer pollutants can destroy or rupture the air sacs, which the body replaces with scar tissue. Destroyed or damaged air sacs or alveoli can not be replaced or repaired. Besides that, the scar tissue often becomes the place where emphysema or cancer begins to develop. It's true that medical treatments can slow or stop the spread of cancer, but medical treatments cannot bring back destroyed or damaged alveoli.

What Do You Think?

Should there be a pollution tax for industries that cause air pollution? How about for cars, lawn mowers, and leaf blowers? How should the money from a pollution tax be spent to improve our atmosphere and to improve the health of the people affected by air pollution?

apply your KNOWLEDGE Suppose you were given the job of monitoring the air pollution levels in your city. What are some of the things you would want to measure?

Smoking

Cigarette smoke is an air pollutant, but it is much more concentrated since it is inhaled directly into the lungs. Like other pollutants in the air, smoke from smoking can cause cancer and damage the alveoli in the lungs. A smoker's lung will look charred or black and bumpy because of the fusion of tiny alveoli into larger air sacs.

So if smoking causes cancer and destroys alveoli, then why do people smoke? Why do some young people start smoking? Why does the Surgeon General keep saying that cigarette smoking is dangerous to your health? Why do you still see ads for smoking in magazines, newspapers, and on billboards? Let's see if we can find out. Here are some interesting statistics to start investigating these questions.

When products are found to be "tainted" or to cause harm, they are usually removed from the market by the manufacturer or by law. Why hasn't this happened with cigarettes?

- The Surgeon General reports that in the United States 390,000 deaths per year are due to smoking.
- About 26% of all adults in this country smoke. In many other countries the percentage is higher.
- In 1985, the total cost of health care associated with smoking was greater than 65 billion dollars. Today it's even more.
- Most smokers start as teenagers. The current starting age averages 16 years. By 17 to 19 years, one-fifth of all teenagers smoke regularly.
- Breathing someone else's exhaled tobacco smoke can be damaging to your lungs.
- Autopsy studies of teenagers show that early, small airway narrowing occurs within the first few years of active smoking.
- Teenage smokers have more colds and coughing symptoms. They also have lower lung vital capacities than do nonsmokers.
- Educating teenagers about smoking, alone, does not prevent them from smoking.
- Less educated people tend to smoke more than high school and college graduates do.

Tobacco contains nicotine and nicotine is a drug. People keep smoking for the nicotine. Nicotine is a stimulant. A stimulant is a substance that makes some people feel more energetic or excited. It is the nicotine that is habit forming, or addictive. The big problem is that nicotine

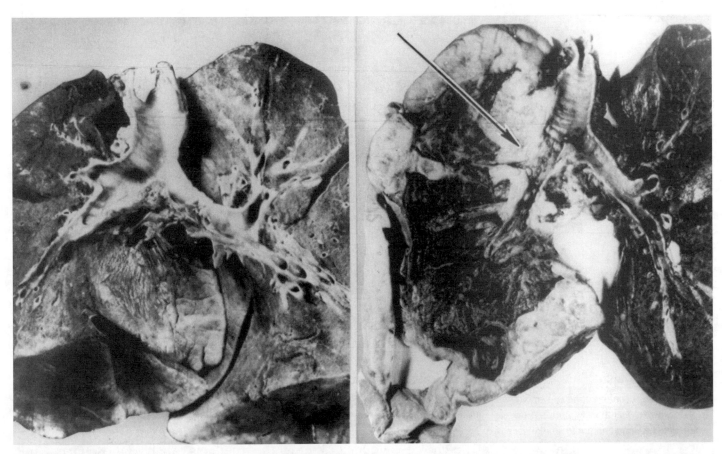

Figure 5.5 The picture on the left shows the lung of a non-smoker. The picture on the right shows a smoker's lung. The arrow points to cancer cells in the smoker's lung.

irritates the lining of blood vessels, which can lead to the formation of plaque. If plaque develops in the coronary arteries, the person can begin to experience symptoms of heart disease. Smoke also irritates your eyes and throat. It stains your teeth, and makes your breath and clothes stink. It makes many people turn away or say nasty things. So why would anyone want to smoke?

Some smokers don't know the risks of smoking. Some feel that bad things happen to other people and not to them. But bad things can happen to you, too. For smokers the biggest risk is permanently damaging the heart and lungs. Smoking can cause death, but smoking causes many problems short of death that can ruin your life. You have probably seen people who get short of breath after only a little exertion. But have you seen people who have to carry around oxygen bottles to help them breathe? Have you seen people who breathe through holes in the front of their necks because they have had throat surgery? In fact, it's because smoking causes so many health problems that the Surgeon General wants everyone to quit. Now, think about why the government cares about you and wants people to quit smoking. Well, one reason is that smokers are sick more often than non-smokers are. When they're sick, they cost everyone a lot of money in paying for health care. Even more importantly, they cause great pain for themselves and their families when they are sick or dying.

MINI ACTIVITY

Motivating through Ads
Collect three or four advertisements about smoking. Study the ads to determine what they tell you. Explain the message of the ad. Who smokes in these ads? What is the life-style these ads are trying to sell to you? To whom do you think the ads are marketing? Use the same kinds of gimmicks these ads use to develop an anti-smoking ad. If you were to create an ad campaign *against* smoking, how could you make it "glamorous" or attractive?

apply your KNOWLEDGE

Who is the Surgeon General? What is the Surgeon General's job in the government?

So what can you make of all you read and hear about smoking? Here are some things to think about:

1. If you smoke you damage your heart, circulation, and lungs. That makes it easier for you to get sick.
2. If you continue to smoke for many years you will probably die of a disease caused by smoking. Some common diseases are lung and throat cancer, heart disease, stroke, and emphysema. The earlier you start smoking, the more likely it is that you will die earlier in your life of one of these diseases.
3. Some people want you to smoke and some want you to quit. Companies making money from cigarettes, advertising, or the sale of tobacco want you to smoke. When anyone buys a pack of cigarettes, cigars, or chewing tobacco, most of their money goes to the tobacco companies.

What Do You Think?

Do you mind if people around you smoke? Explain why you feel the way you do. How would you ask someone not to smoke near you?

Anyone who is interested in your health does not want you to smoke because you will be sick, die early, and cost the health care system lots of money.

The choice is yours. Gather all the true facts. Analyze those facts. Then choose wisely. But be sure to make your own choice! Be sure to think carefully before you choose.

Activity 5-1
Smoke in Your Lungs

Introduction
What does smoking do to your lungs? In this activity you build a smoking machine, then explore how cigarettes put nicotine, tar, and oils in a smoker's lungs.

Materials
1-liter Clear, plastic detergent or soda bottle
Cotton balls, enough to fill the bottle
Clay or play dough
Rubber tubing or straw (to fit over cigarette)
Activity Report

Procedure
Step 1   Fill a bottle with cotton balls.

Step 2   Surround a short piece of rubber tubing with clay or playdough and place it in the neck of the bottle.

Step 3   Place a cigarette into the tube (outside the classroom).

Step 4   Have your teacher light the cigarette (outside the classroom).

Step 5   "Inhale" and "exhale" the bottle slowly by squeezing and relaxing your grip.

Step 6   Observe and record what happens.

Step 7   Clean up as directed by your teacher.

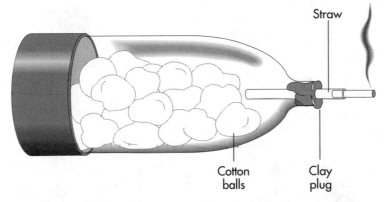

Figure 5.6 This drawing illustrates a smoking machine.

Emphysema
Emphysema is a disease of smokers that destroys the delicate membranes of the air sacs. As a result, individual alveoli combine into larger and larger air sacs. Remember Activity 2-2 in which you calculated surface areas of golf balls, tennis balls, and volleyballs? Think back to the comparison between the gas exchange capacity of the golf balls with the volleyball. What do you think the fusion of small alveoli into larger and larger air sacs does to the gas exchange capacity of the lungs? Another problem is the development of scar tissue. As the alveoli are damaged, scar tissue forms and the lungs become less elastic. That means the lungs cannot stretch as much when the diaphragm contracts. So the vital capacity of the lungs goes down even more. The reduced diffusion capacity and reduced vital capacity makes it very hard for an emphysema patient to breathe.

Activity 5-2
Emphysema

Introduction
In emphysema, the surface area of all the alveoli is reduced. This reduction of surface area makes it difficult for the patient to breathe. In this activity you model the reduction of surface area of alveoli in emphysema.

Materials
Microscope
Slides of normal lung tissue
Slides of lung tissue from a person with emphysema
Your teacher may substitute photographs of each.
   (But the magnification of each must be the same.)
Calculator
Tennis balls
Golf balls
4-liter freezer bags
Balloon
Lung models from Activity 1-1
Activity Report

Procedure
Step 1   Observe the differences between normal lung tissue slides and emphysema slides. Answer questions 1 through 3 on the Activity Report.

Step 2   Refer to *Activity 1-1: How Do You Breathe?*

a. Pair up with another team. One team should put another balloon inside the existing balloon.
b. Each team should repeat step 5 from *Activity 1-1: How do You Breathe?*
• Step 5. Use the tied off neck of the balloon as a handle and pull down. Hold for a second. Let go.
c. Compare the expansion of the single balloon with the expansion of the double (two balloons).

Step 3   Refer to *Activity 2-2: The More the Airier* procedure 3.
• Calculate the surface areas of the golf ball, tennis ball, and volleyball. If you don't remember how to calculate the surface area check Steps 1 to 3 of Activity 2-2: *The More the Airier.*
a. Fill the bag with golf balls (air sacs).
b. Remove half of the golf balls and replace them with tennis balls.
c. Compare the surface area of all the golf balls with the surface area created by half golf balls and half tennis balls together.

Step 4   Answer the questions on the Activity Report.

Choking
Now think back to the beginning of this unit when we described the air-ways to the lungs. What happens when someone chokes? Suppose you're watching someone eat lunch. Suddenly the person grabs at his throat and can't talk or breathe. The choker looks scared and starts to pass out, turning a pale blue. First think. Then act. If there is anyone else around send that person to get medical help. The possibility is that food has probably stuck over the windpipe around the epiglottis. The person can't talk, because the food is blocking the vocal cords. The person can't breathe because the airway is blocked. You know you must act because the person's lungs are losing oxygen to the blood. No new oxygen is getting in, but the blood is taking up what oxygen is there from the air sacs. The amount of carbon dioxide in the air sacs is rising. When the person tries to breathe, the effort causes the food to be sucked down even harder. This blocks the choker's airway even more.

The first thing to do in a choking situation is to get emergency medical help. Someone trained in the Heimlich maneuver (also called the Obstructed Airway maneuver) can perform the procedure to dislodge the object. The Heimlich maneuver is a first aid technique that helps someone who is choking. This technique is different for children than it is for adults, so it is important to get the correct training that teaches the procedure. A healthcare professional such as the school nurse, a doctor, or an Emergency Medical Technician (EMT) can demonstrate how to do the Heimlich maneuver safely.

Keeping Your Lungs Healthy

It's easy to take for granted everything that your lungs do for you. Think about how hard you have to work to breathe when you're congested with a cold or the flu. Life feels pretty uncomfortable when breathing is hard to do. Many factors can affect your lungs that you can't control, such as getting viral infections. But there are a lot of other factors you can control that help keep your lungs healthy. The following are some things you can control that will help keep your lungs healthy.

Good nutrition: A good diet can provide the energy, vitamins, and minerals your body needs. Remember, for example, that the hemoglobin in your red blood cells delivers oxygen from your lungs to your body's cells. What is hemoglobin? Hemoglobin is protein that needs iron to work. Therefore, you need to get iron in your diet. Foods rich in iron include leafy vegetables, dried fruits, and red meats.

Sleep: Have you ever noticed that when you haven't gotten enough sleep you get sick more easily? Your immune system directly benefits from lots of sleep. Sometimes bacteria or viruses sneak by your cilia and mucus escalator. If you are well rested, your immune system will be better able to fight off illness.

Water/fluids: Remember that your air passages moisten, warm, and clean incoming air. They do this through their natural structure and through the production of mucus. Keeping these tissue cells moist helps them to function more efficiently. Bacteria and viruses can get into your system more easily if your air passages are dry. When you drink a lot of fluids, you also improve your circulation. When you improve your circulation, you help flush out (get rid of) bacteria and viruses.

Air quality: Air pollution can lead to many lung problems, from the common cold to cancer. What can you do? Live and work in places with good air quality. But since that's not always possible, you must become more aware about the things you and your community do that affect the air around you. For example, smoking pollutes the air around you, as does driving a car, using small gasoline motors such as lawn mowers and outboard motors, having a fire in the fireplace, and using spray cans. (Cleaners, paints, and deodorants often come in spray cans.)

MINI ACTIVITY

Advertising for Good Health

Create an ad campaign for your school promoting good health and healthy lungs. Determine what people should and should not do in order to promote good health and healthy lungs. Design a flyer or brochure that provides the necessary information to distribute to your friends and families. Include diagrams and art work. BE CREATIVE! It is sometimes difficult to tell people things they do not want to know so be persuasive. The brochure should be informative, convincing, appealing, and neat.

Exercise: Getting regular exercise keeps your body strong. Exercise improves the strength and capacity of your heart and lungs. The stronger all your body systems are, the better they can fight disease.

Alcohol/Drugs: Alcohol and drugs change how your cells function. They frequently cause dehydration, which makes you feel thirsty. When your airways are dry, more bacteria and viruses can get by the natural barriers. More importantly, drugs and alcohol affect your judgment. Poor judgment can lead to poor decisions that negatively affect your health. For example, activities such as sniffing paint or glue fumes destroys tissue in your airways and reduces the protective mucus lining—not to mention destroying cells in your brain and liver. Seeing someone on a mechanical lung or artificial respirator because of sniffing glue helps us see the results that poor decisions can have on your respiratory system.

Review Questions

1. Why is mucus important to your health?

2. How do viruses and bacteria spread? What are the best ways to protect yourself from getting them?

3. What is the difference between pneumonia, bronchitis, asthma, and emphysema?

4. Describe at least four reasons why smoking is hazardous to your health.

5. Describe three ways you can help your lungs stay healthy.

Glossary

aerosol   chemicals suspended in the air like a mist or a fog.

air pollution   a name for the substances put into the air that contaminate our atmosphere.

alveoli   air sacs that are clustered around the smallest bronchioles. They are the site of gas exchange between the air and the blood.

arteries   blood vessels that carry blood from the heart.

asthma   a condition in which the small airways leading to the air sacs get narrow.

bacteria   tiny organisms that can enter your body and change the function of cells causing sickness.

blood gas levels   the amount of carbon dioxide and oxygen in your blood.

bronchi   tubes that branch off the trachea and lead to the smaller bronchioles.

bronchioles   tubes that branch off the bronchi and repeatedly branch into smaller and smaller tubes.

bronchitis   infection of the upper airways and smaller air tubes. Two symptoms are wheezing and a persistent cough.

capillaries   tiny blood vessels through which materials pass between cells and the blood.

carbon dioxide   a waste gas your cells make, exhaled when you breathe.

cellular respiration   a process in which the cell uses oxygen to break down fuel molecules to produce energy and the waste products—carbon dioxide and water.

cilia   tiny hairs that line airways for breathing.

control systems   systems in your body that help regulate and maintain body functions at a normal level.

diaphragm   a large dome-shaped sheet of muscle, located beneath the lungs.

diffusion   the natural movement of particles from an area of high concentration to an area of low concentration.

emphysema   a disease that destroys the delicate membranes of the air sacs. As a result, individual alveoli combine into larger and larger air sacs.

epiglottis   a flap-like trap door that closes when you eat or drink to keep food out of the airways.

esophagus   the tube that carries what you eat and drink to your stomach.

feedback information   information used by control systems indicating a change has occurred and a function is deviating from normal.

glottis   a slit-like opening between your vocal cords.

glucose   a simple sugar molecule.

larynx   the voice box—connects the pharynx with the windpipe that goes to the lungs.

lungs   a pair of sponge-like organs responsible for the exchange of gases during breathing.

mucus   a secretion produced by tissues in the airways that help moisten air and protect your body from particles and infection.

mucus escalator   term for the way particles are moved towards the throat. The motion of the cilia combined with the mucus create the mucus escalator.

nervous system   the system composed of the brain, spinal cord, and nerve cells that connect to all parts of the body.

oxygen   a tasteless, odorless, and colorless element of air. About 20% of air you breathe is oxygen.

pharynx   the throat—the section of the digestive system that leads from the mouth to the esophagus.

photosynthesis   the process in which plant cells use carbon dioxide to produce sugars (or starches) from energy (sun).

pneumonia   condition when the small air tubes and air sacs in the lungs get infected with bacteria. The infected cells produce fluids, which makes breathing difficult.

thermostat   a device which controls temperature. The "thermostat" in the human body is located in the part of the brain called the hypothalamus.

trachea   the main breathing tube to the lungs.

tuberculosis (TB)   one of the ten leading causes of death in the world. Active tuberculosis bacteria destroy the air sacs in a person's lungs.

veins   blood vessels that carry blood to the heart.

venules   small blood vessels that carry the blood from the capillaries to the larger veins.

viruses   tiny parasites smaller than bacteria.

vital capacity   the biggest possible breath a person can breathe in and exhale out.